Physiological Aspects of Imprinting and Homing Migration in Salmon:

Emerging Research and Opportunities

Hiroshi Ueda
Hokkaido University, Japan

A volume in the Advances in Environmental Engineering and Green Technologies (AEEGT) Book Series

IGI Global
DISSEMINATOR OF KNOWLEDGE

Published in the United States of America by
IGI Global
Engineering Science Reference (an imprint of IGI Global)
701 E. Chocolate Avenue
Hershey PA, USA 17033
Tel: 717-533-8845
Fax: 717-533-8661
E-mail: cust@igi-global.com
Web site: http://www.igi-global.com

Library of Congress Cataloging-in-Publication Data

Names: Ueda, Hiroshi, 1951- author.
Title: Physiological aspects of imprinting and homing migration in salmon : emerging research and opportunities / by Hiroshi Ueda.
Description: Hershey, PA : Engineering Science Reference, [2020] | Includes bibliographical references and index. | Summary: "This book examines the physiological mechanisms of natal stream imprinting and homing in salmon with special references to hormone, olfaction, memory, and behavior"-- Provided by publisher.
Identifiers: LCCN 2019036924 (print) | LCCN 2019036925 (ebook) | ISBN 9781799820543 (hardcover) | ISBN 9781799820550 (paperback) | ISBN 9781799820567 (ebook)
Subjects: LCSH: Salmon--Homing. | Imprinting (Psychology) | Salmon--Migration. | Salmon--Behavior.
Classification: LCC QL638.S2 U428 2020 (print) | LCC QL638.S2 (ebook) | DDC 597.5/61568--dc23
LC record available at https://lccn.loc.gov/2019036924
LC ebook record available at https://lccn.loc.gov/2019036925

This book is published in the IGI Global book series Advances in Environmental Engineering and Green Technologies (AEEGT) (ISSN: 2326-9162; eISSN: 2326-9170)

British Cataloguing in Publication Data
A Cataloguing in Publication record for this book is available from the British Library.

Advances in Environmental Engineering and Green Technologies (AEEGT) Book Series

ISSN:2326-9162
EISSN:2326-9170

MISSION

Growing awareness and an increased focus on environmental issues such as climate change, energy use, and loss of non-renewable resources have brought about a greater need for research that provides potential solutions to these problems. Research in environmental science and engineering continues to play a vital role in uncovering new opportunities for a "green" future.

The **Advances in Environmental Engineering and Green Technologies (AEEGT)** book series is a mouthpiece for research in all aspects of environmental science, earth science, and green initiatives. This series supports the ongoing research in this field through publishing books that discuss topics within environmental engineering or that deal with the interdisciplinary field of green technologies.

COVERAGE

- Radioactive Waste Treatment
- Alternative Power Sources
- Cleantech
- Waste Management
- Policies Involving Green Technologies and Environmental Engineering
- Electric Vehicles
- Green Technology
- Water Supply and Treatment
- Green Transportation
- Renewable Energy

IGI Global is currently accepting manuscripts for publication within this series. To submit a proposal for a volume in this series, please contact our Acquisition Editors at Acquisitions@igi-global.com or visit: http://www.igi-global.com/publish/.

The Advances in Environmental Engineering and Green Technologies (AEEGT) Book Series (ISSN 2326-9162) is published by IGI Global, 701 E. Chocolate Avenue, Hershey, PA 17033-1240, USA, www.igi-global.com. This series is composed of titles available for purchase individually; each title is edited to be contextually exclusive from any other title within the series. For pricing and ordering information please visit http://www.igi-global.com/book-series/advances-environmental-engineering-green-technologies/73679. Postmaster: Send all address changes to above address.

Titles in this Series

For a list of additional titles in this series, please visit:
http://www.igi-global.com/book-series/advances-environmental-engineering-green-technologies/73679

Artificial Intelligence and IoT-Based Technologies for Sustainable Farming and Smart Agriculture
Pradeep Tomar (Gautam Buddha University, India) and Gurjit Kaur (Delhi Technological University, India)
Engineering Science Reference • © 2020 • 350pp • H/C (ISBN: 9781799817222) • US $215.00

Predicting, Monitoring, and Assessing Forest Fire Dangers and Risks
Nikolay Viktorovich Baranovskiy (National Research Tomsk Polytechnic University, Russia)
Engineering Science Reference • © 2020 • 417pp • H/C (ISBN: 9781799818670) • US $235.00

Effects of Emerging Chemical Contaminants on Water Resources and Environmental Health
Victor Shikuku (Kaimosi Friends University College, Kenya)
Engineering Science Reference • © 2020 • 350pp • H/C (ISBN: 9781799818717) • US $225.00

Waste Management Techniques for Improved Environmental and Public Health Emerging Research and Opportunities
Sang-Bing Tsai (Zhongshan Institute, University of Electronic Science and Technology of China, China & Wuyi University, China) Zhengxi Yuan (Zhongshan Institute, University of Electronic Science and Technology of China, China) Jian Yu (Research Center for Environment and Sustainable Development of China Civil Aviation, Civil Aviation University of China, China) and Xuexin Liu (Capital University of Economics and Business, China)

For an entire list of titles in this series, please visit:
http://www.igi-global.com/book-series/advances-environmental-engineering-green-technologies/73679

IGI Global
DISSEMINATOR OF KNOWLEDGE

701 East Chocolate Avenue, Hershey, PA 17033, USA
Tel: 717-533-8845 x100 • Fax: 717-533-8661
E-Mail: cust@igi-global.com • www.igi-global.com

Table of Contents

Foreword

"He wrote the book on that!" is one of the highest compliments anyone can receive in a discussion of any subject. The phrase is colloquial, but the message is clear. It might not literally mean that the individual wrote a book, but it certainly does mean that the individual is a key expert on that subject. Such individuals define the subject, develop ideas, provide critical information and present the definitive perspective. However, in this case, the statement is both figuratively and literally correct. Hiroshi Ueda is the key expert on the study of the imprinting and homing migration of Pacific salmon, and he is the author of the book on this subject. What else can I say?

It is a distinct personal privilege and pleasure to add my Foreword to this remarkable book. I have been truly fortunate to know Hiroshi as a great personal friend and research colleague for most of my professional career. He and his research group have hosted my visits to Japan on numerous occasions, and like salmon, he has migrated across the North Pacific Ocean to spend time in our research group in Oregon. He has contributed immensely to our own efforts to study salmon behavior, and we are indebted to him for that. This book shows how much everyone owes him.

The more I study the behavior of salmon, the more I am impressed with their incredible abilities. Everyone should take the opportunity to observe wild adult salmon during their spawning migration. No matter how often I witness this behavior, it never fails to increase my respect for the fish. Of course, I can physically hold a salmon and remember that perhaps 5 years ago I held the parents of the fish in my hands in exactly the same location at almost exactly the same time of the year. That in itself is incredible. During the intervening years in the life of this fish, it has perhaps swum a total distance equivalent to half the breadth of the Pacific Ocean without prior experience or guidance from its parents to reach an appropriate foraging location in the Ocean, after which it has returned to this precise location to complete its life cycle. This fact never ceases to amaze me. Of course, indigenous peoples from around the North Pacific Ocean have always had great respect for salmon and what they provide to their culture and societies. It is impossible to live anywhere around the North Pacific Ocean and not be influenced by salmon. In addition, salmon are unequalled as subjects for scientific study. For better or worse, they are the most nearly perfect fish.

I was once asked to summarize the importance of Pacific salmon at a large public conference on the management and conservation of freshwater systems in the Pacific Northwest. I was allotted 5 minutes for my presentation. Of course, the task is remarkably simple, and I had time for questions and discussion. We can summarize the importance of Pacific salmon in three words: anadromous, philopatric, and semelparous. They migrate between freshwater and the ocean and return to spawn in freshwater. They return to the location where their parents spawned to carry out their own spawning behavior, and they do this only once. The fact that we can convey so much in so few words speaks volumes – as Hiroshi has done in this book. The importance of Pacific salmon in terms of human history, culture, economics, politics and societal values cannot be overstated. We can imagine a world without Steller's sea cow (*Hydrodamalis gigas*), the great auk (*Pinguinus impennis*), the passenger pigeon (*Ectopistes migratorius*) or even the Chinese sturgeon (*Psephurus gladius*) as a result of human activities, but we continue to go to extraordinary lengths to preserve Pacific salmon. We cannot imagine the Pacific Northwest without salmon.

The great value that we place upon Pacific salmon, together with the extraordinary complexity of their life cycle and behavior, has attracted much scientific attention. Pacific salmon are known from every biology textbook, every account of the history of people in this region and every television program on migration behavior. Virtually any scientist can find Pacific salmon to be a rich subject for study, ranging from research on molecular genetics to decadal oceanic population cycles.

In this book, Hiroshi Ueda takes us on a guided personal tour of what must be the most spectacular and intriguing components of their behavior:

their imprinting and homing migration. This book captures every aspect that makes the subject so fascinating, including history, basic science, human social values, economics, conservation and, perhaps most importantly, our individual values. Of course, science is fundamental to all of these aspects. One of the most telling parts of this book is the extensive list of collaborators, supervisors, students and scientific colleagues that Hiroshi acknowledges. These extensive connections occur among all scientists, as science can be described as one of the most truly collaborative human activities. It also tells us a great deal about Hiroshi himself.

Perhaps it is not stretching the comparison too much here to compare the interactions and specialized contributions of so many scientific colleagues in our research field to what might be involved in the imprinting and homing behavior of salmon. Salmon exhibit an intricate, complex sequence of ecological and life history patterns that all must occur in remarkable coordination. It is because Hiroshi has done so much of this research himself that he can lead us through the complexity of what appears to be the simple, direct life cycle of the salmon. Starting with the fundamental questions of phylogeny, Hiroshi proceeds to build an increasingly complex account of the life history of Pacific salmon that culminates in the key topics of this book. Exactly how is it that salmon can find their way during their extended movements without the benefit of prior experience? In doing so, he provides us with the answer that in a very real sense, salmon have evolved to answer this fundamental question and complete their life cycle. Hiroshi describes the progress in our science, including where we started, how we have progressed and where we now find ourselves. In doing so, he clearly defines the scientific protocol for the future, including the important remaining questions and what it will take to answer them. Thus, he has developed and proposed a program for future research on salmon migration and homing behavior. It is appropriate that he ends his book considering the production and harvesting of salmon, for these activities are inextricably intertwined with our understanding of salmon biology and life history.

Of course, this will remain the definitive book on the subject for the foreseeable future. In part, this is because so much information on the subject has now been revealed, with the disparate bits of information being fitted together, and the subject is now being presented by the central figure in this field of science. When I try to help people understand the kinds of research questions we are addressing, I ask them to think like a salmon. Often, a change in perspective can provide new insights, especially if that perspective is one that is very different from our own. I also tell them that what we are trying to do in our research is to ask the salmon a simple question that it can understand and then tell us a simple answer that we can understand in turn. It is remarkably easy to make science complicated. We see examples of this around us almost every day. It is easy to be beguiled by the power of large numbers, complex experimental designs, powerful computers and intricate statistical analyses. However, complexity seldom provides real understanding. It is extremely difficult to keep things simple, but it is remarkably rewarding. The challenge is to ask a simple question that cuts to the essence of the matter, so that the answer can only be a simple yes or an equally satisfying no. How many times have we heard a lecture or read a paper or book and declared to ourselves, "Of course, that is so obvious!", but we did not see it ourselves. It is reported that when Thomas Huxley first read The Origin of Species by Darwin he remarked "How extremely stupid not to have thought of that!", a comment that is truly remarkable given that Huxley was undoubtedly one of the most knowledgeable scientists of his time. We should all thank Hiroshi for taking a very extensive body of complex information and refining it to reach a conclusion that we can all understand.

David L. G. Noakes

Fisheries and Wildlife and Oregon Hatchery Research Center, Oregon State University, Corvallis, USA

Foreword

From birds flying over the Mediterranean Sea between Europe and Africa or Arctic terns flying from one end of the planet to the other to monarch butterflies flying over North America to spend the winter in a valley in Mexico, sea turtles swimming to remote islands to breed, and humpback whales swimming from Hawai'i to Alaska, the migrations of animals have fascinated humans since time immemorial. What cues in the environment tell them it is time to leave, what clues guide them, how do conditions along the route accelerate or delay them, and how have the patterns evolved? Along with these and many other animal species, Pacific and Atlantic salmon provide classic examples of migratory behavior. After spawning in the cool waters of streams and lakes, they feed and grow for a few months or years in freshwater before migrating to sea. They then spend several years in the open waters of the Atlantic Ocean or Pacific Ocean before commencing their homeward migration to breed in the specific stream that they left years earlier.

The fact that salmon home to natal sites for breeding has been known for the last century, but salmon have resisted the efforts of scientists to pry into their minds and bodies; thus, many questions remain about the migrations of these remarkable fishes. There are different species of salmon (and their close relatives, trout and char), each with its own variations on the overall themes of migration and homing. How can the key activities of the life cycle be accomplished despite so many different specific patterns? Most of these fish migrate to sea in the spring, but some do so in the fall; adults of one population or another return from the ocean in every month of the year and spawn in every month of the year. What hormones and other physiological processes trigger the migration to the sea and back? What roles do temperature and day length play in these remarkable movements accomplished by hundreds of millions of adults and billions of juveniles each year? How do they find their way around the open ocean, and back again to the coastal region where they started? How do they know which stream they left out of the thousands of small streams and tributaries of large branching rivers?

After decades of research into the patterns and processes of salmon migration and homing, Dr. Hiroshi Ueda has provided us with the benefits of his experimental work, and he has also reviewed that of colleagues from around the world, all in an attempt to reveal the mysteries of salmon. Like the peeling of an onion, we see the multiple layers of complexity underlying the behavior patterns involved in downstream migration, dispersal at sea, return migration, and the identification of odors learned early in life, yet remembered without reinforcement years later, that distinguish the home stream from all others. The lengthening of days triggers changes in the endocrine glands that set in motion a series of hormonal events preparing salmon for their transition between fresh and salt water, stimulating them to migrate and spurring the memory of odors unique to their home stream. They must learn these chemical cues, sometimes referred to as a bouquet of odors, and seek them out years later as another series of endocrine events occurs to reverse the process.

The work that Dr. Ueda presents is international in scope, including the many experiments that he and his colleagues have conducted in Japan and research conducted elsewhere by other scientists. We learn, among other things, of his many experiments on the importance of amino acids as a class of chemicals that help identify the home stream and the intricate biochemical processes involved in olfactory learning and memory. The movements of individual salmon in the open ocean and in lakes in Japan are also reported in detail based on various biotelemetry techniques. Indeed, Dr. Ueda conducted experiments with salmon in these lakes to test hypotheses generated from observations of salmon at sea. In addition to these aspects of the behavior and physiology of salmon, Dr. Ueda presents information on the importance of salmon in commercial fisheries and the rapid rise of salmon aquaculture in recent decades. In this regard, salmon differ from some of the other organisms that serve as models for the study of migration, such as migratory birds and butterflies, or the study of evolution, such as guppies and finches.

In preparing this book, Dr. Ueda has done a major service to the scientific community by bringing together so much information from so many sources and consolidating the great body of work that he and his students have conducted. The mysteries of salmon migration are not entirely solved, but this important new book will help guide future generations of scientists interested in salmon migrations and homing to decide what questions to pose and how to tackle them.

Thomas P. Quinn

School of Aquatic and Fishery Sciences, University of Washington, Seattle, USA

Preface

Salmon are among the most well-known and most commonly eaten fish and represent one of the most important fishery resources in the world. They are cold-water fish species and used to exist only in the Northern Hemisphere from middle to high latitudes. However, they were transplanted to the Southern Hemisphere from the 1870s to 1900s, and they inhabit now cold-water regions around the world. They are born and die in freshwater but can live in both freshwater and seawater and migrate between rivers and oceans, showing amazing abilities to precisely home to their natal stream. However, their dynamic life cycles and mysterious natal stream imprinting and homing migration abilities are not well understood.

In the prehistoric period, it has been speculated that peoples living near rivers where salmon exist might have been aware of the upstream migration of salmon to their spawning grounds in autumn and might prefer to live in ancient dwellings near salmon spawning grounds because of the possibility of easily obtaining salmon body mass as a valuable source of autumn and winter food. Ancient houses of the Ainu people, an indigenous people in Hokkaido, Japan, were located near salmon spawning grounds in rivers. The Eskimo people, an indigenous people in Alaska, USA, refer to chum salmon dog salmon because they use chum salmon body mass as bait for dogs, which are important partner animals for dog sledding.

In the Western countries, the first description of salmon homing was provided in the book "The Complete Angler" by Izaak Walton in 1653 in England: "Sir Francis Bacon observed, the age of a salmon exceeds not ten years, much of this has been observed by tying a ribbon or some known tape or thread, in the tail of some young salmons, which have been taken in weirs as they have swimmed toward the salt water, and then by taking a part of them again with the known mark at the same place at their return from the sea, which is usually about six months after, which has inclined many to think, that every salmon usually returns to the same river to which it was bred." Extensive efforts had been made to examine why salmon can home to their natal streams; however, these attempts were long unable to resolve the homing mechanisms of salmon.

Arthur Hasler and Warren Wisby reported in 1951 that in Wisconsin, USA, "bluntnose minnow were able to discriminate successfully between chemical differences of two Wisconsin creeks after two months' training, preliminary tests with salmon (no species description) proved that they can detect the stream odors and that they were able to discriminate between them, and it is postulated that the nature of the guiding odor must be such that it have meaning only for those salmon conditioned to it during their freshwater sojourn." Furthermore, they reported in 1954 that "Silver salmon (coho salmon) captured in traps in Issaquah Creek and its East Fork. The olfactory pits of approximately one-half of the fish were occluded before all of them were displaced to the Issaquah approximately one mile below the junction of the two streams. The control salmon, unlike those with occluded olfactory pits, showed an excellent ability to repeat their original choice at the stream juncture. This is interpreted to mean that the operated fish were not able to differentiate between the two streams but were distributing themselves in a random fashion. The results, therefore, are in accord with those which would be expected if the fish were relying on their sense of smell in making this choice." Following their report on the olfactory imprinting and homing mechanisms of salmon, a large number of extensive studies on the properties of natal stream odors, how juvenile salmon and adult salmon remember and retrieve natal stream odors and what kinds of sensory systems salmon use in open-water migration were conducted in western countries. However, there are still many unsolved mysteries regarding salmon imprinting and homing migration to their natal streams.

The olfactory hypothesis is sufficient to explain short-distance migration within the natal stream; however, it is almost impossible to explain the long-distance migration of salmon in the open water. The geomagnetic imprinting hypothesis proposes that salmon may migrate from their ocean habitats to their natal streams in a narrow migration corridor along a magnetic field isoline (Bracis & Anderson, 2012). The collective navigation hypothesis has been proposed for homeward migration in anadromous salmon, which suggests that they use geomagnetic, celestial, and olfactory cues (Berdahl et al., 2014). An inherited magnetic map has been proposed as a navigational solution that enables salmon to migrate to oceanic foraging grounds without prior experience or guidance from older conspecifics (Putman, 2015). Population-specific strategies (Byron & Burke, 2014) and social interactions (Berdahl et al., 2017) have also been demonstrated as mechanisms of oceanic spawning migration in anadromous salmon.

In Japan, the homing ability of chum salmon was recognized in 1763 by Buheiji Aoto, who established a system for protecting homing salmon in the River Miomote, in Niigata, creating artificial channels known as "Tanegawa" where adult salmon could spawn, and all freshwater fisheries were prohibited in spring when juvenile salmon undertake their downstream migration. Artificial salmon propagation techniques were introduced from Western countries in the late 1870s, and an artificial chum salmon propagation system involving the release of juveniles from hatcheries into their natal streams and catching adults in their natal area several years later began to be implemented in the late 1880s (Ueda, 2015). In Japan, children in the early grades of elementary school in the northern area now release the hatchery-originating juvenile chum salmon into streams in spring while saying "Please come back to the stream as big adults". Children of the later grades are happy to know that adult salmon home to their natal streams in autumn several years later. Salmon provide useful teaching materials for students in elementary schools. Most Japanese chum salmon juveniles, with a body weight of approximately 1 g, are released from many hatcheries into streams in spring, then achieve seawater adaptability around the mouth of the stream and begin their feeding migration along the coast of Hokkaido, after which they undergo long-distance feeding migration in the ocean around the Sea of Ohotsuku, Bering Sea, and Gulf of Alaska for several years. In the Bering Sea in summer several years later, the onset of gonadal maturation is the trigger for the onset of the homing migration of adult salmon. They reach coastal areas of northern Japan, where most of these fish are captured by setting fish nets. A few fortunate adults that escape the fish nets can enter their natal stream, where they are captured by wire fishing. They are then transported to hatcheries where artificial insemination is performed to produce offspring, and the fish then die.

I have studied one of the most important and interesting mysteries of biological and fisheries science: the imprinting and homing migration of salmon, for more than 40 years, mainly using four anadromous two landlocked Pacific salmon populations in Hokkaido, Japan (Ueda, 2004, 2011, 2012, 2014, 2015, 2016, 2018, 2019; Ueda & Tsukamoto, 2014; Ueda et al., 2000, 2007). This research has been conducted with many excellent hard-working students, graduate students, postdoctoral fellows, and talented experts among my Japanese and international colleagues. In this book, I would first like to introduce salmon species from around the world and their dynamic and complicated life cycle associated with fish migration and climate change. Second, I present the most recent findings regarding the physiological mechanisms of natal stream imprinting and homing in salmon with special reference to hormones, olfaction, memory, and behavior. Finally, I describe salmon resources including salmon commercial fisheries, aquaculture and propagation systems around the world.

I believe that the scientific frontiers of the 21st century are space science, marine science and brain science. Space science research may be difficult; however, marine science and brain science research using salmon as the main research materials may not be difficult. I sincerely hope that this book will be useful to the general public and students with interest in salmon as well as scientists carrying out environmental and physiological research who seek to understand the mechanisms of the migration of aquatic animal species such as salmon.

REFERENCES

Berdahl, A., Westley, P. A. H., Levin, S. A., Couzin, I. D., & Quinn, T. P. (2014). A collective navigation hypothesis for homeward migration in anadromous salmonids. *Fish and Fisheries*, *17*(2), 525–542. doi:10.1111/faf.12084

Berdahl, A., Westly, P. A. H., & Quinn, T. P. (2017). Social interactions shape the timing of spawning migrations in an anadromous fish. *Animal Behaviour*, *126*, 221–229. doi:10.1016/j.anbehav.2017.01.020

Bracis, C., & Anderson, J. J. (2012). An investigation of the geomagnetic imprinting hypothesis for salmon. *Fisheries Oceanography*, *21*, 2–3. doi:10.1111/j.1365-2419.2012.00617.x

Byron, C. J., & Burke, B. J. (2014). Salmon ocean migration models suggest a variety of population-specific strategies. *Reviews in Fish Biology and Fisheries*, *24*(3), 737–756. doi:10.1007/s11160-014-9343-0

Hasler, A. D., & Wisby, W. J. (1951). Discrimination of stream odours by fishes and relation to parent stream behaviour. *American Naturalist*, *85*(823), 223–238. doi:10.1086/281672

Putman, N. F. (2015). Inherited magnetic maps in salmon and the roles of geomagnetic change. *Integrative and Comparative Biology*, *55*(3), 396–405. doi:10.1093/icb/icv020 PubMed

Ueda, H. (2004). Recent biotelemetry research on lacustrine salmon homing migration. Memoirs of National Institute of Polar Research, 58(SI), 80–88.

Ueda, H. (2011). Physiological mechanisms of homing migration in Pacific salmon from behavioral to molecular biological approaches. *General and Comparative Endocrinology*, *170*(2), 222–232. doi:10.1016/j.ygcen.2010.02.003 PubMed

Ueda, H. (2012). Physiological mechanisms of imprinting and homing migration in Pacific salmon Oncorhynchus spp. *Journal of Fish Biology*, *81*(2), 543–558. doi:10.1111/j.1095-8649.2012.03354.x PubMed

Ueda, H. (2014). Homing ability and migration success in Pacific salmon: Mechanistic insights from biotelemetry, endocrinology, and neurophysiology. *Marine Ecology Progress Series*, *496*, 219–232. doi:10.3354/meps10636

Ueda, H. (2015). The current propagation system and physiological studies of Japanese chum salmon. *Fisheries Science*, *81*(2), 219–228. doi:10.1007/s12562-014-0847-2

Ueda, H. (2016). Physiological mechanisms of imprinting and homing migration of Pacific salmon. *Aqua-BioScience Monographs*, *9*(1), 1–27. doi:10.5047/absm.2016.00901.0001

Ueda, H. (2018). Migration and Navigation in Fish. In M. K. Skinner (Ed.), Encyclopedia of Reproduction. (2nd ed., Vol. 6, pp. 84–89). Elsevier; doi:10.1016/B978-0-12-809633-8.20540-2.

Ueda, H. (2019). Sensory mechanisms of natal stream imprinting and homing in Oncorhynchus spp. *Journal of Fish Biology*, *95*(1), 293–33. doi:10.1111/jfb.13775 PubMed

Ueda, H., Leonard, J. B. K., & Naito, Y. (2000). Physiological biotelemetry research on the homing migration of salmonid fishes. In A. Moore & I. Russell (Eds.), *Advances in Fish Telemetry* (pp. 89–97). Lowestoft: Crown Copyright.

Ueda, H., & Tsukamoto, K. (Eds.). (2014). *Physiology and Ecology of Fish Migration*. Boca Raton, London, New York: CRC Press.

Ueda, H., Yamamoto, Y., & Hino, H. (2007). Physiological mechanisms of homing ability in sockeye salmon: From behavior to molecules using a lacustrine model. *American Fisheries Society Symposium*, *54*, 5–16.

Walton, I. (1653). *The Complete Angler*. London: Richard Marriot.

Wisby, W. J., & Hasler, A. D. (1954). Effect of olfactory occlusion on migrating silver salmon (Oncorhynchus kisutch). *Journal of the Fisheries Research Board of Canada*, *11*(4), 472–478. doi:10.1139/f54-031

Acknowledgment

The reason that I started my physiological research on salmon was both positively and negatively influenced by my late father, Dr. Meiichi Ueda, who was an ecologist of wild mice and the director of Wild Mouse Research Laboratory of the National Forestry Institute, Hokkaido, Japan. The positive influence was that he took me fishing frequently in my childhood, and this experience caused fish to become my favorite wild animals. The negative influence was that when I was a high school student, he took me to the forest field where wild mice were trapped and taught me how to weight the gonads of wild mice that were preserved in formalin. I asked him "Why should I weight the gonads of wild mice, and what kind of biology do these procedures belong to?". He answered clearly that "The gonad weight of wild mice can predict the birth number of juvenile mice and I could forecast how wild mice would cause damages to forest, and this work belongs to the field of ecology". I liked the biology and wild animals very much but completely despised these procedures and could not touch or observe the worm blood of animals after this experience. I decided to never study ecology in my research carrier. However, I liked and respected my father's serious attitude toward biology and decided to study salmon, with cold blood, mainly using physiological approaches, but my work gradually began to incorporate with ecological approaches.

First, I would like to give my special thanks to my four mentors: the late Prof. Kiichiro Yamamoto, late Prof. Hiroya Takahashi, Prof. Yoshitaka Nagahama, and Prof. Kohei Yamauchi, who taught me how to study the physiology of salmon as an under graduate student of the Faculty of Fisheries, Hokkaido University, a graduate student of Graduate School of Fisheries Science, Hokkaido University, a postdoctoral fellow of National

Institute for Basic Biology, a faculty member of Faculty of Fisheries, Hokkaido University, respectively.

I would also like to express my sincere gratitude those individuals who have guided me during my research carrier: Prof. Teruaki Uchida of Kyushu University who introduced me to the faculty position I obtained at the of University of Occupational and Environmental Health; Prof. Sunao Fujimoto, who was my supervisor at the University of Occupational and Environmental Health; Dr. Peter F. Hall, my supervisor at the Worcester Foundation of Experimental Biology, MA, USA; Prof. Abraham A.L. Kierszenbaum, supervisor at the University of North Carolina, NC, USA; and Prof. Akihiko Hara, who guided me my faculty position of Hokkaido University.

My heartfelt gratitude goes to those individuals who have carried out valuable collaborative research in Japan: the late Prof Haruo Kanatani, Dr. Arimune Munakata, Dr. Takashi Kitahashi, Prof. Akihisa Urano, Prof. Masafumi Amano, Prof. Munehiko Iwata, Prof. Kazuyoshi Tsutsui, Prof. Shinji Adachi, Prof. Hironori Ando, Prof. Toyoji Kaneko, Prof. Masayuki Iigo, Dr. Masatoshi Ban, Dr. Hideaki Yamada, and Dr. Hiroaki Chiba for their salmon hormone research; Prof. Takayuki Shoji, Prof. Kenzo Kurihara, Prof. Norio Suzuki, Prof. Masakane Yamashita, Prof. Hideaki Shibata, Prof. Masahiko Watanabe, Dr. Ikuhiro Kida, Dr. Toshiaki Hara, Dr. Takashi Denboh, Dr. Kohtaro Konno, Dr. Shigeho Ijiri, and Mr. Hiroyuki Haruna for their salmon olfaction and memory research; Prof. Masahide Kaeriyama, Prof. Katsumi Tsukamoto, Dr. Hideji Tanaka, Prof. Yasuhiko Naito, Prof. Hitoshi Araki, Dr. Shigehiko Urawa, Dr. Masa-aki Fukuwaka, Dr. Kinya Nagata, Dr. Shunpei Sato, Dr. Hiroshi Kawamura, and Dr. Mitsuhiro Nagata for their salmon behavioral research; and Prof. Takashi Matsuishi, Dr. Ikutaro Shimizu, Mr. Nobuhisa Koide, Mr. Naoyuki Misaka, Mr. Tomoichiro Tanaka, Mr. Noriaki Takahashi, Mr. Katsuya Nakao, Mr. Makoto Fujii, Mrs. Yukiyo Iimura, and Mr. Yoshihide Shimakawa for their salmon resource studies; and those who have conducted such research from around the world: Mr. Steve Anglea, Prof. Terence Barry, late Prof. Howard Bern, Dr. Nolan Bett, Prof. John Caprio, Dr. Bill Chang, Prof. John Cockrem, Dr. Steven Cooke, Dr. Nancy Davis, Dr. Andrew Dittman, Mr. Yong Huang, Prof. Scott Hinch, Dr. Steve

McCormick, Dr. Andy Moore, Prof. David Noakes, Dr. Nathan Putman, Dr. Mee-Ra Rhyu, Prof. Thomas Quinn, Prof. Carl Schreck, Dr. Penny Swanson, Dr. Villy Tan-Fermin, Prof. B. Senthilkumaran, Dr. Vehanen Teppo, Dr. Carol Woody, Prof. Graham Young, and Prof. Yonathan Zohar. In particular, many thanks to Prof. David Noakes and Prof. Thomas Quinn for writing the Forward of this book.

My heartfelt thanks also go to those who have studied in my laboratory as graduate and Ph.D. students: Dr. Hiroyuki Sakano, Dr. Hisaya Nii, Dr. Yuya Makiguchi, Dr. Kazufumi Hayashida, Ms. Ayako Sato, Mr. Kiyoshi Orito, Mr. Eiji Fujiwara, Ms. Yukiko Matsushita, Mr. Yuki Nabatake, Mr. Masumi Akita, Ms. Emiko Nakagawa, Mr. Masaya Nomura, Ms. Kanna Tagita, Mr. Yoshifumi Konno, Mr. Yuki Yoshida, and Mr. Koji Miyoshi for behavioral and ecological researches; Prof. Hideaki Kudo, Dr. Ramji Kumar Bhandari, Dr. Minwyelet Mingist Zerihun, Dr. Masahiro Fukaya, Ms. Kumiko Shimozawa, Mr. Shinya Taniyama, Mr. Kenji Seo, Mr. Shingo Matsumoto, Ms. Hanae Iwase, Ms. Tomoto Kitani, Mr. Nobuhisa Kozu, Mr. Shih-Pin Lee, Mr. Kosuke Fukaya, Ms. Hiroko Watanabe, Ms. Hanako Shimazuka, Mr. Shigeo Tsuchida, Mr. Naohiro Furukawa, Mr. Reiichi Murakami, Mr Shingo Nakamura, Mr. Taro Nakamura, and Mr. Kanji Makise for endocrinological and physiological researches; and Dr. Munetaka Shimizu, Dr. Ibrahim Fares, Dr. Koji Sato, Dr. Hiroshi Hino, Dr. Yuzo Yamamoto, Dr. Nina Ileva, Dr. Hiroshi Bando, Mr. Shinyo Iwasaki, Mr. Satoshi Yanagi, Mr. Koji Saito, Mr. Daisuke Nishikawa, Ms. Fumi Morinishi, Mr. Go Miyasaka, Ms. Mayuka Noma, Mr. Keita Kiya, Ms. Ayumi Nakasa, Mr. Kohei Baba, Ms. Sayaka Ishizawa, Mr. Eisuke Itagaki, Mr. Kohei Baba, Ms. Harumi Nakano, Ms. Kanako Okada, Mr. Takashi Okubo, Mr. Naoki Katayama, Ms. Kaoru Inada, and Mr. Shunsuke Hakamada for olfactory research, and as postdoctoral fellows and visiting scientists: Prof. Jill Leonard, Dr. Jason Sandahl, Dr. Nathan Miles, Dr. Arjan P. Palstra, Prof. Jin Deuk-Hee, Prof. Cheol Young Choi, Mr. Marcus Welker, Mr. Joseph Lemanski, and Ms. Katharine Self.

Much of the data from my laboratory presented in this book have been published in original and review manuscripts by many journals: *American Fisheries Society Symposium, Aqua BioScience Monographs, Aquaculture, Aquaculture Science, Canadian Journal of Fisheries and Aquatic Sciences,*

Cell and Tissue Research, Chemical Senses, Comparative Biochemistry and Physiology Part B, Cybium, Ecology of Freshwater Fish, Fisheries Management and Ecology, Fisheries Science, General and Comparative Endocrinology, Journal of Fish Biology, Journal of Neurochemistry, Marine Ecology Progress Series, Memoirs of National Institute of Polar Research, North Pacific Anadromous Fish Commission Bulletin, PLoS ONE, Scientific Report, Zoological Science, and *Zoological Studies*, and books; *Advances in Fish Telemetry, Encyclopedia of Reproduction. 2nd Edition*, and *Physiology and Ecology of Fish Migration*. I acknowledge these journals and books.

My sincere appreciation toes to the following institutes: the Hokkaido National Fisheries Research Institute, the Hokkaido Research Organization Salmon and Freshwater Fisheries Research Institute, the Hokkaido Aquaculture Promotion Corporation, the Hokkaido Salmon Propagation Association, the Chitose Salmon Aquarium, the Shibetsu Salmon Museum, and the Sapporo Salmon Museum for providing Pacific salmon, research opportunities, and research facilities. Special thanks to the Hokkaido Aquaculture Promotion Corporation for the publication of this book.

My salmon research was supported mainly by a Grant-in-Aid for Scientific Research from the Ministry of Education, Culture, Sports, Science and Technology of Japan (grant number 07660227, 08660216, 09460082, 11556035, 18208017, 23658151, 23380106), the Japan Society for the Promotion of Science (grant number P97207, P07764, S11097, L11557, SP12062, SP14035, SP16057), the Program for Revitalization Promotion from the Japan Science and Technology Agency, the Hokkaido Foundation for the Promotion of Scientific and Industrial Technology, the Mitsubishi Foundation, the Mitsui & Co. Ltd, the Akiyama Life Science Foundation, and Hokkaido University.

Finally, I dedicate this book to my family: my late parents, Meiichi and Masako, my loveliest wife, Keiko, and my two sons and their families, Yuichi, Ayumi, Rio, Saya, and Yoshiki, Mayumi, Seiya, and Reika, whose love and support made it possible to publish this book and my life in salmon research fruitful.

Chapter 1
Salmon Species and Life Cycles

ABSTRACT

Salmon are very interesting fish with complicated life cycles. They are born and die in freshwater but can live in both freshwater and seawater and migrate between rivers and oceans. Salmon previously inhabited only cold-water regions of the Northern Hemisphere but have been transplanted to the Southern Hemisphere and are now distributed around the world, and these fish have an amazing ability to imprint and precisely home to their natal stream. This chapter describes the terminology and taxonomy of salmon and related species associated with their phylogenetic evolution, the salmon species found around the world, the life cycles of Pacific salmon with special reference to imprinting and homing migration, the migration of other important migratory fish species, and the impact of climate change on migratory fish.

DOI: 10.4018/978-1-7998-2054-3.ch001

INTRODUCTION

Salmon are among the most well-known and appreciated fish species in the world, but the terminology related to these fish is quite complicated. Salmon and related species are collectively termed salmonids and are characterized by paired pelvic fins in the middle of the abdomen and an adipose fin on their back near the caudal fin and by spawning in freshwater. They are born and die in freshwater but can live in both freshwater and seawater, migrating between rivers and oceans. Salmon have an amazing ability to remember their natal streams and retrieve that information. However, their mysterious abilities to undergo imprinting in their natal stream during their downstream migration, leading to seawater adaptation, as juveniles and to home to their natal stream during their upstream migration for reproduction as adults are not well known. The amazing imprinting and homing migration abilities of salmon were recognized long ago. However, it is not well understood how salmon evolved from their ancestral species, where salmon exist around the world, and when salmon undergo imprinting and home to their natal stream.

Salmon and related species are classified by phylogenetic analyses into three major subfamilies and ten genera according to the identification of lineages considered to be ancestrally anadromous and those that have retained the ancestral state of pure freshwater residency. Salmon are mainly divided into species showing iteroparity (Atlantic salmon; genus *Salmo*) or semelparity (Pacific salmon; genus *Oncorhynchus*). However, some species have been transferred from *Salmo* to *Oncorhynchus* according to recent morphological and genetic studies. Salmon used to exist only in the Northern Hemisphere, but some species have been transplanted to the Southern Hemisphere and now they inhabit cold water regions around the world. The life cycles of Pacific salmon differ among species depending on the timing of juvenile downstream migration and adult upstream migration. Salmon are well known for their migratory abilities, but there are many migratory fish species, and most of them are valuable fisheries resources. Recent serious unpredictable climate change has caused many negative and a few positive impacts on migratory fish species.

Terminology and Taxonomy of Salmon and Related Species

The terminology and taxonomy of salmon and related species are quite complicated. These fish belong to the scientific classification of order Salmoniformes and family Salmonidae and are collective termed salmonids. The family Salmonidae comprises three subfamilies: Thymallinae, Coregoninae, and Salmoninae. There are ten genera in these subfamilies: *Thymallus* (Graylings), *Stenodus* (Beloribista), *Prosopium* (Round whitefishes), *Coregonus* (Whitefishes), *Brachymystax* (Lenoks), *Hucho* (Siberian taimen), *Salmo* (Atlantic salmon), *Parahucho* (Japanese huchen or Sakhalin taimen), *Salvelinus* (including *Salvethymus*: Chars), and *Oncorhynchus* (Pacific salmon) (Crête-Lafrenière et al., 2012; Shedko et al., 2013).Recently reported mitochondrial cytochrome oxidase subunit I (*COI*) gene sequences, along with partial amino acid sequences, demonstrated the origin of salmonids from an ancestor closely related to modern *Salvelinus* species (Artamonova et al., 2018).

Salmonids have been estimated to have diverged no later than 88 million years ago, when a whole-genome duplication occurred in the ancestral salmonid causing the diploid ancestor to become tetraploid, after which 40-50 million years passed before the subfamilies diverged (Macqueen & Johnston, 2014). The sister species of salmonids have been indicated to be the freshwater fish belonging to *Esox* (Pike fish) on the basis of whole-mitochondrial genome sequences, suggesting that salmonids originated from freshwater species (Campbell et al., 2013). Some salmonids undergo a downstream migration to the ocean, where they grow and undertake an upstream migration in freshwater for reproduction; this kind of migration is known as anadromous migration (anadromy: sea-run form). Some salmonids spend their whole life in rivers or lakes, which is referred to as freshwater residency (resident form). Although the conception of the phylogeny of Salmonidae differs among researchers, the analysis of the temporal evolution salmonid fish lineages based on a mitogenome tree has revealed the phylogenetic relationships of the three subfamilies and ten genera of salmonids, revealing lineages considered to be ancestrally anadromous and to have retained the ancestral state of pure freshwater residency (Figure 1).

Figure 1. Phylogenetic profile of three major subfamily and ten genera of salmonidae. Modified from Macqueen & Johnston (2014). A and F show lineages considered to be ancestrally anadromy and to have retained the ancestral state of pure freshwater residency, respectively.

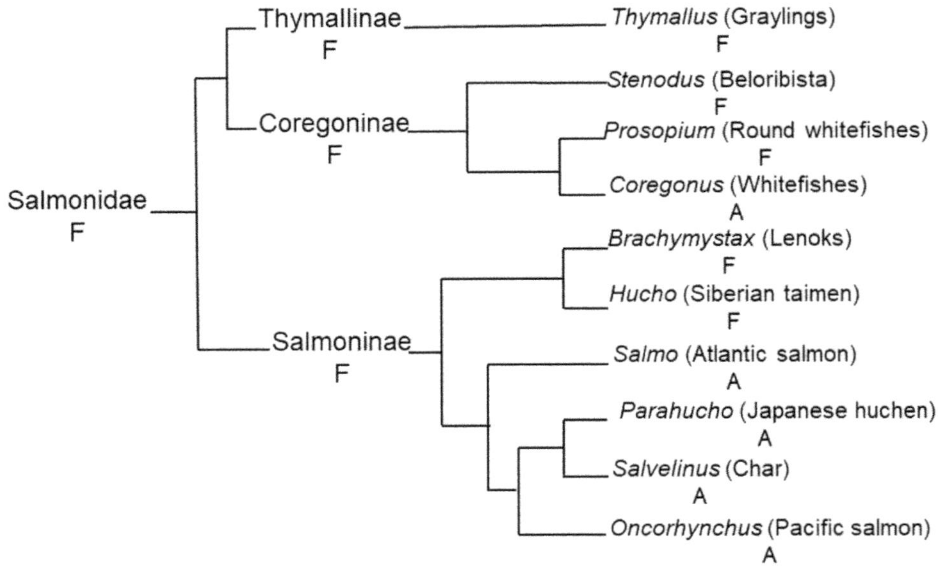

Salmon Species Around the World

There are eight species of Pacific salmon: pink or humpback salmon (*O. gorbuscha*), chum or dog salmon (*O. keta*), sockeye or red salmon (*O. nerka*), Chinook or king salmon (*O. tschawytscha*), coho or silver salmon (*O. kisutch*), masu or cherry salmon (*O. masou*), cutthroat trout (*O. clarkii*), and rainbow trout or steelhead (*O. mykiss*), which are resident and anadromous forms, respectively. The Atlantic salmon consist of two species: Atlantic salmon (*S. salar*) and brown or sea trout (*S. trutta*). All Pacific salmon except for rainbow trout (steelhead) and cutthroat trout are semelparous species in which all adults die after reproducing only once during their lifetime (semelparity). Atlantic salmon, rainbow trout (steelhead) and cutthroat trout are iteroparous species in which some adults known as kelt produce offspring several times during their lifetime (iteroparity). Rainbow trout (steelhead) and cutthroat trout used to be considered Atlantic salmon species (genus *Salmo*) and were transferred to the Pacific salmon (genus *Oncorhynchus*) in 1989 on the basis of morphological and genetic research (Smith & Stearley, 1989).

It can be difficult to explain how salmon diverged into semelparous Pacific salmon and iteroparous Atlantic salmon during salmonid evolution. One bold speculation might be put forth in relation to the glacial period. Around the Atlantic Ocean, mainland northern Europe and northeastern America were heavily covered with thick ice sheets at this time. Atlantic salmon are expected to have increased their reproductive chance by repeated reproduction due to the serious decrease in the chance of upstream migration. In contrast, around the Pacific Ocean, west Eurasia and northwestern America were moderately covered in thick ice sheets. Pacific salmon did not necessarily increase their reproductive chance in a similar way due to the many chances for upstream migration. The mechanisms underlying the differences between semelparous Pacific salmon and iteroparous Atlantic salmon are scientific mysteries that should be solved in the future.

Atlantic and Pacific salmon used to exist only in the Northern Hemisphere from middle to high latitudes. However, rainbow trout and brown trout were transplanted into rivers and lakes of Chile and New Zealand around the 1870s. Moreover, fertilized eggs of Chinook salmon from the Sacramento River, California, USA, were transplanted into the Waitaki River in New Zealand between 1901 and 1907; they were able to colonize the area so that anadromous Chinook salmon could be produced in the Southern Hemisphere (McDowall, 1994; Quinn et al., 2001). In contrast, fertilized eggs of chum salmon were transplanted from Japan to Chile between 1974 and 1986, but they could not achieve successful colonization (Japan International Cooperation Agency, 1991). The main difference between successful colonization in New Zealand and unsuccessful colonization in Chile might be that New Zealand is an island nation, while the coast of Chile is characterized by steep cliff. In New Zealand, Chinook salmon can migrate to find their natal stream via either clockwise or counterclockwise routes because it is an isolated island. However, in Chile, chum salmon could not find a homing route due to the steep cliffs.

Pink and chum salmon are widely dispersed in the North Pacific Ocean. Sockeye salmon inhabit the eastern areas and high-latitude western areas of the North Pacific Ocean. In Japan, anadromous sockeye salmon were last found in 1900 in eastern Hokkaido and have not been observed since. In contrast, the resident form, lacustrine sockeye salmon, which are known as

Figure 2. Life history of two different types of Pacific salmonid species. Dotted line: chum and pink salmon; Solid line: Sockeye, Chinook, Coho and Masu salmon.

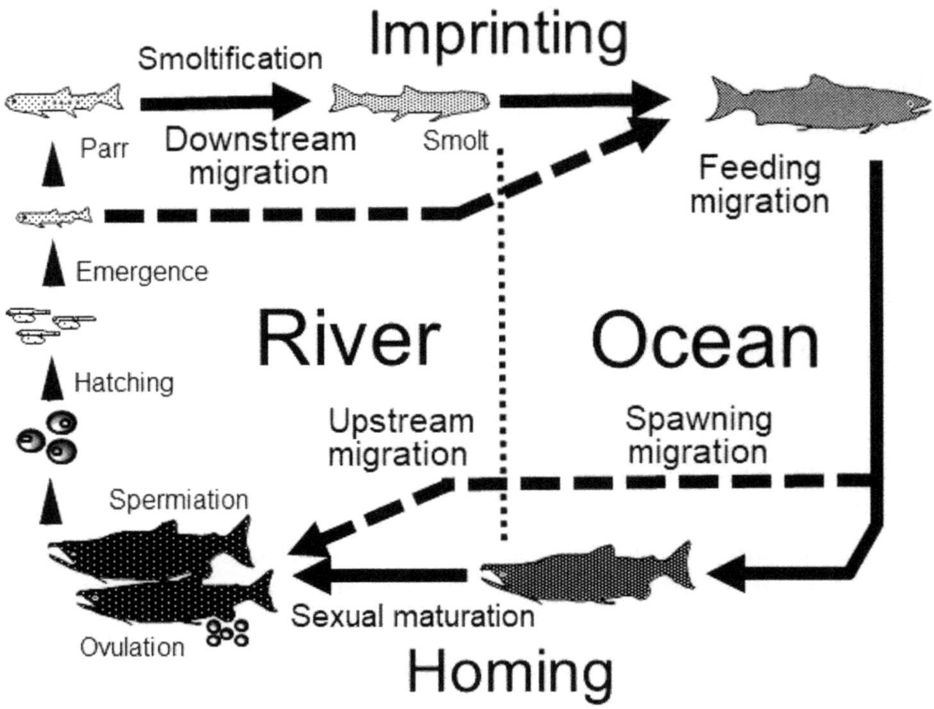

kokanee in North America, survived in Lake Akan and were transplanted to and colonized many lakes in northern Japan. Japan's endemic species "Kunimasu" (Black kokanee: *O. kawamurae*), was thought to have become extinct due to lake acidification in Lake Tazawa, Akita Prefecture. However, Kunimasu were discovered in Lake Saiko, Yamanashi Prefecture, owing to transplantation from Lake Tazawa around the 1930s (Nakabo et al., 2011). Chinook, coho salmon and steelhead are distributed mainly in eastern areas, and masu salmon are distributed only in western areas of the North Pacific Ocean.

In the North America, there are several subspecies of cutthroat trout (*O. clarkii*): along the Pacific coast, coastal cutthroat trout (*O. c. clarkii*) and Lake Crescent cutthroat trout (*O. c. crescenti*); in the Great Basin, Alvord cutthroat trout (*O. c. alvordensis*), Bonneville cutthroat trout (*O. c. utah*), Humboldt cutthroat trout (*O. c. humboldtensis*), Lahontan cutthroat

Figure 3. Phylogenetic division of seven Pacific salmon using retropositional genome analyses (A) as modified from Murata et al. (1996). Biomass that shows the average total catch by species from 1952 to 1975 and the ocean distribution index that indicates the matrix numbers of an area of 2 degrees latitude × 5 degrees longitude where salmon were caught until 1985 (B) as modified from Kaeriyama & Ueda (1998). Four Pacific salmon in Japan are framed.

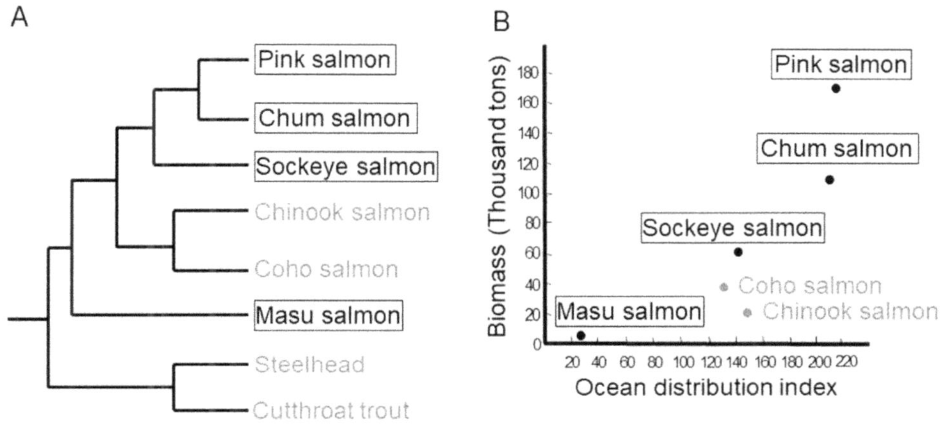

trout (*O. c. henshawi*), and Paiute cutthroat trout (*O. c. seleniris*); in the northern Rockies, Snake river fine-spotted cutthroat trout (*O. c. behnkei*), Westslope cutthroat trout (*O. c. levisi*), and Yellowfinn cutthroat trout (*O. c. maddonaidi*); and in the southern Rockies, Colorado River cutthroat trout (*O. c. pleuriticus*), Greenback cutthroat trout (*O. c. stomias*), and Rio Grande cutthroat trout (*O. c. virginalis*).

In Japan, there are several subspecies of masu salmon: "Satsukimasu" (red-spotted masu salmon. *O. m. ishikawae*) is an anadromous form distributed in southern Japan. "Biwa trout" (*O. m. rhodurus*) and "Amago salmon" (*O. rhodurus*) are lacustrine and resident forms of Satsukimasu, respectively. In Taiwan, Formosan land-locked salmon (*O. m. formosanus*) are the most southerly distributed salmon in the Northern Hemisphere because they are survivors from the glacial period. These fish are subspecies of masu salmon and inhabit streams at an 1,800 m elevation in Shei-Pa National Park in central Taiwan.

Life Cycles of Pacific Salmon

In Pacific salmon, all underyearling juveniles of pink salmon and chum salmon undergo downstream migration in spring within a few months after emergence, and there is no resident form. Adult pink salmon and chum salmon undergo upstream migration in autumn within a few weeks of spermiation and ovulation after 2 years and 2-7 years of oceanic feeding migration, respectively. In contrast, juveniles of sockeye salmon, Chinook salmon, coho salmon, and masu salmon remain in freshwater rivers or lakes for several years, undergo smoltification (parr-smolt transformation: PST), in which the physiological changes allowing seawater adaptability to occur, and then undertake downstream migration in spring. There are resident individuals who spend their whole life in rivers or lakes. Adults undergo upstream migration after several years of oceanic feeding migration in autumn at least several months prior to spermiation and ovulation (Figure 2). One exception is the so-called winter Chinook salmon in the Sacramento River, California, USA, which undergo upstream migration in winter and spawn in spring (Quinn, 2018).

Retropositional genome analyses of these six Pacific salmon indicated that pink salmon and cutthroat trout might be the most advanced and primitive salmon species, respectively (Murata et al., 1996) (Figure 3A). Pink salmon are also the most widely distributed species in the North Pacific Ocean and exhibit the largest population size, while masu salmon appear to show the most restricted distribution in the West Pacific Ocean and the smallest population (Kaeriyama and Ueda 1998) (Figure 3B). Although the homing accuracy of these salmon has not been compared in detail, it is believed that masu salmon return to their natal stream with the highest precision and that pink salmon are more likely to stray into a nonnatal stream. If most salmon show highly accurate homing to the natal stream, there will be little chance of increasing their distribution area or their population size. Additionally, they might encounter the dangerous possibility of a reduction of their genetic diversity. The relationship between salmon evolution and homing accuracy is one of the most interesting questions from the viewpoint of salmon evolution.

FISH MIGRATION

In addition to salmon, there are a variety of other migratory fish species that show interesting and mysterious life cycles. There are approximately 30,000 species of fish around the world, and only 165 species are categorized as migratory fish. Migratory fish acquire many important biological functions that make their successful migration possible, and most of them are valuable fisheries resources. Either alone or in schools, they undergo drift migration from their spawning habitat to their nursery habitat for recruitment; undergo feeding migration from their nursery habitat to their feeding habitat for growth; and undergo spawning migration from their feeding habitat to their spawning habitat for reproduction (Harden-Jones, 1968). In special cases, drift migration occurs again from the spawning habitat to the feeding habitat if adults survive after reproduction as Atlantic salmon kelt (Figure 4). The fish continuously perform a migration loop to produce the next generations, which allows them not only to exploit different food resources and adapt to environmental changes but also to regulate their population density and widen their distribution. Fish

Figure 4. Fish migration showing migration loop that drifting migration for recruitment from spawning habitat to nursery habitat, feeding migration for growth form nursery habitat to feeding habitat, spawning migration for reproduction from feeding habitat to spawning habitat, and post-reproductive movement from spawning habitat to feeding habitat as modified from Harden-Jones (1968) and Ueda (2018).

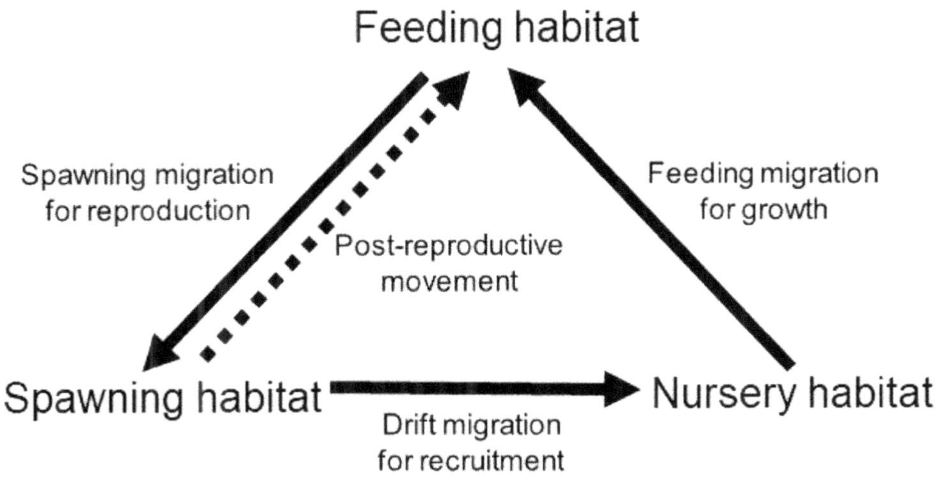

Figure 5. Diadromous fish migration showing catadromous species (eel), anadromous species (salmon), and amphidromous seawater species (Japanese sea bass), and freshwater species (ayu) as modified from Ueda (2018).

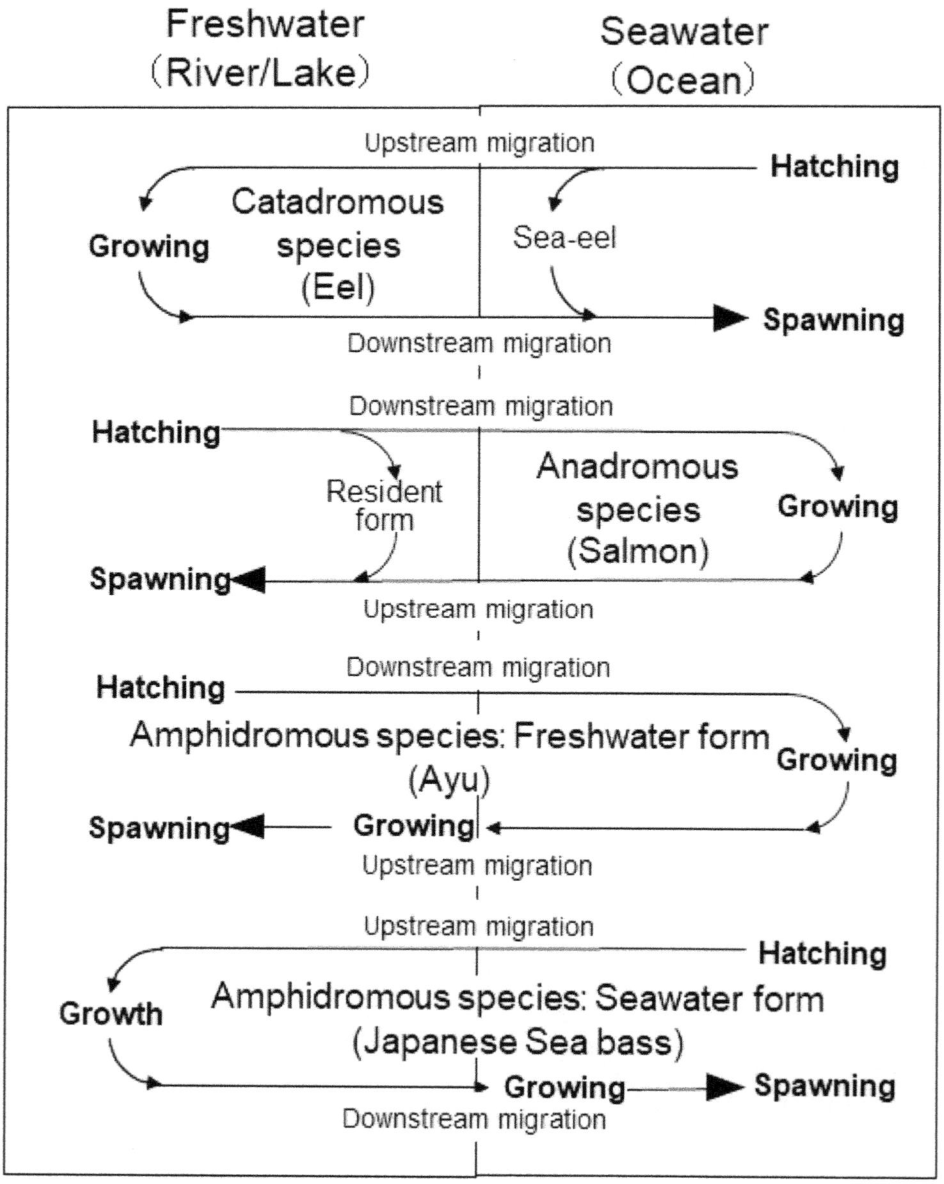

migration is classified into diadromous migration between freshwater and seawater and nondiadromous migration in only either freshwater or seawater (Ueda & Tsukamoto, 2014; Ueda, 2018).

Diadromous Migration

Diadromous migration is divided into catadromous migration, anadromous migration and amphidromous migration. These different types of migration are characterized by hatching, growing, and spawning habitats in either freshwater or seawater (Fig. 5). Catadromous and anadromous migrations start and end in seawater and freshwater, respectively. Amphidromous migration is a special type migration in which both catadromous and anadromous migrations occur, and these fish are further divided into seawater and freshwater forms that can grow in both freshwater and seawater (Table 1). The appearance frequency of diadromous migration is reported to correlate with primary productivity depending on latitude. In low-latitude warm water areas, primary productivity is low in seawater and high in freshwater, and the number of catadromous migrating species increases. In high-latitude cold-water areas, primary productivity is high in seawater

Table 1. Classification and representative species of migratory fish

I. Diadromous species		
	1. Catadromous species: Eel	
	2. Anadromous species: Salmon; Sturgeon; American shad; Striped bass	
		1) Sea-run form
		2) Reident form
	3. Amphidromous species	
		1) Seawater species: Japanese sea bass
		2) Freshwater species: Ayu (Sweet fish)
II. Non-diadromous species		
	1. Oceanodromous species	
		1) Highly migratory species: Tuna
		2) Coastal migratory species: Herring
	2. Potamodromous species	
		1) Fluvial species: Mekong giant catfish
		2) Lacustrine species: Isaza

and low in freshwater, and the number of anadromous migrating species increases. In addition, increases. In mid-latitude moderate-temperature areas, the number of amphidromous migrating species increase (Gross et al., 1988).

Catadromous Migration

Eels (genus *Anguilla*) are worm-water fish of 16 species (Aoyama, 2009). As like salmon, eels are divided mainly into Atlantic eel (American eel: *A. rostrate* and European eel: *A. anguilla*) and Pacific eel (Japanese eel: *A. japonica* and short-finned eel: *A. australis*). They are born in the ocean, and their larval "leptocephalus" undergoes transplantation by drifting with the aid of the ocean current towards estuaries on the continental shelf, where they metamorphose into a transparent larval stage known as "glass eels". Many glass eels undergo upstream migration into freshwater habitats, where they remain for 5-20 years and grow into "yellow eels", and some of these fish become "sea-eels" that never enter freshwater spending their entire lifetime in the ocean (Tsukamoto et al., 1998). Eels become "silver eels" at the time of the onset of gonadal maturation and undergo downstream migration. Adult eels undergo oceanic spawning migration to the spawning area, and all adults die after reproduction (semelparity).

The spawning areas of European eels was found in the Sargasso Sea in the 1920s, but many mysteries regarding juvenile and adult migration between the Sargasso Sea and freshwater habitats distributed in European countries remain unsolved. Juveniles are reported to possess a magnetic map of the North Atlantic Ocean and to target the Gulf Stream system (Naisbett-Jones et al., 2017). The spawning migration of adults has been observed via biotelemetry, revealing that they exhibit diel vertical migrations to deeper water during the day and shallower water at night at swimming speeds of 3~ 47 km/day (Righton et al., 2016).

The spawning area of the Japanese eels was discovered in west of the Marian Islands, where adults and newly hatched larvae are caught during new moon periods throughout the spawning season indicating that spawning occurs in shallower layers of 150~200m (Tsukamoto et al., 2011). However, there are many mysteries regarding how adult eels migrate and navigate in the ocean and find their spawning sites, including what kinds of sensory systems are involved.

Anadromous Migration

In addition to Pacific and Atlantic salmon, sturgeon (four genera: *Acipenser*, *Huso*, *Scaphirhynchus*, and *Psedoscaphirhynchus*), American shad (*Alosa sapidissima*), and striped bass (*Morone saxatilis*) are anadromous species. These species all spawn in freshwater, undergo downstream migration into seawater where they grow, and make upstream migration for reproduction.

Amphidromous Migration

Japanese sea bass (*Lateolabrax japonicus*) is a seawater form that exhibits amphidromous migration. This species is native to the West Pacific Ocean, and spawns in coastal seas in winter. Juveniles inhabit seawater, brackish water, and freshwater where they grow from spring to autumn and undergo upstream migration in rivers to avoid high water temperatures. Adults undergo downstream migration to the coastal seas for reproduction.

Ayu, or sweetfish (*Plecoglossus altivelis*), is a freshwater form that exhibits amphidromous migration, and spawns in rivers in autumn. Juveniles undergo drift migration downstream to the coastal seas, where they feed on plankton in winter. Young fish undergo upstream migration in lower and middle reaches, where they feed on attached algae, spawn in autumn, and die after reproduction (semelparity).

Nondiadromous Migration

Nondiadromous migration is divided into oceanodromous migration, which is further divided into highly migratory and coastal migratory forms, and potamodromous migration, which is further divided into fluvial and lacustrine forms.

The highly migratory form of oceanodromous migration is observed in true tuna species (genus *Thunnus*) and other tuna species (four genera: *Allothunnus*, *Auxis*, *Eurhynnus* and *Katsuwonus*). Tunas are endothermic species that can maintain the temperature of certain parts of their body (skeletal muscles, eyes, and brain) above the temperature of ambient seawater, which supports faster swimming speeds (up to 75 km/h) and survival in cooler ocean environments. They are distributed throughout the oceans of the world, generally in tropical and temperate waters at latitudes ranging between approximately 45° north and south of the equator. They carry out migrations of significant distances across oceans for feeding and reproduction. They are iteroparous species, and their relatively large spawning areas in the ocean differ among species.

The coastal migratory form of oceanodromous migration is observed in herring, which are divided into three species: Arucanian herring (*Clupea bentincki*), Atlantic herring (*C. harengus*), and Pacific herring (*C. pallasii*). They migrate in schools from the open ocean to the coastal sea in intertidal and subtidal environments and spawn in a variety of seawater environments during different seasons among different species. Their eggs are laid on the seabed (rock, stones, gravel, or sand) or submerged vegetation (eelgrass or seaweed) and adhere to gravel, seaweed, or stones in layered clumps.

The fluvial form of potamodromous migration is observed in Mekong giant catfish (*Pangasianodon gigas*), which is a critically endangered species native to the Mekong basin. These fish are recognized as the world's largest freshwater species, weighing up to 350 kg. Although little is known about their migrations in the Mekong River, they congregate during the beginning of the rainy season and migrate upstream to spawn in the Mekong River.

The lacustrine form of potamodromous migration is observed in isaza (*Gymnogobius isaza*), which is an endemic species in Lake Biwa, Japan. These fish are a freshwater pelagic species that inhabits offshore areas of the lake and migrate to the shore to spawn under stones in spring. Males care for the eggs in their nest.

IMPACTS OF CLIMATE CHANGE ON MIGRATORY FISH

Recent unpredictable serious climate change affecting oceans, rivers, lake levels, thermal structure, ocean currents and acidification (seawater and freshwater) has caused many negative and a few positive impacts on important fisheries resources and habitats and the distribution of migratory fish. Migratory fish belonging to cold-water species and warm-water species show different responses to water temperature-dependent climate change.

The stocks of the cold-water species Japanese chum salmon peaked in 1996 because of the successful improvement of propagation techniques and favorable habitat conditions in the North Pacific Ocean. However, the number of adult returns has decreased rapidly in recent years. By using the sea surface temperature (SST) database for the North Pacific Ocean,

Kaeriyama & Urabe (2018) analyzed the effects of potential global warming impacts on the growth and trophic level of Japanese chum salmon and found that the areas of optimum temperature (8-12 °C) for chum salmon have decreased markedly in the Sea of Okhotsk and the Gulf of Alaska and that the area of optimum temperature in July, when juveniles migrate to the Sea of Okhotsk, has shifted northward from Hokkaido Island. These results suggest that global warming has profound effects on the migration route, growth and survival of Japanese chum salmon populations in the North Pacific Ocean.

In contrast, the stocks of the warm-water species Japanese amberjack, or yellow tail (*Seriola quinqueradiata*), have increased steadily, mainly due to high-water temperatures along their migratory routes in recent years. Their distribution areas have shifted to the north, and their catches have increased in northern Japan, especially in Hokkaido (Ueda, 2018).

Ocean and freshwater ecosystems may potentially be affected by climate change-related acidification. CO_2-induced freshwater acidification caused reductions in growth and damages olfactory function in the early developmental stage of pink salmon (Ou et al, 2015). In addition, elevated seawater CO_2 affects coho salmon olfactory-mediated behavior, neural signaling, and gene expression within the peripheral and central olfactory systems, and it has been reported that elevated CO2 causes significant behavioral impairment, likely driven by alterations in higher-order neural signal processing within the olfactory bulb (Williams et al., 2018).

CONCLUSION

Salmon and related species are collectively termed salmonids and are characterized distinctive pelvic and adipose fins and by spawning in freshwater. These fish are recognized as having originated from freshwater species, but some of them have evolved to undergo downstream migration to the ocean (most likely due to the high primary productivity of the ocean and the low primary productivity of rivers at high latitudes) and to undergo upstream migration into rivers for reproduction, thus showing anadromous migration. Atlantic salmon (genus *Salmo*) are iteroparous species in which some adults known as kelt produce offspring several times during their lifetime (iteroparity). All Pacific salmon (genus *Oncorhynchus*) except for rainbow trout (steelhead) and cutthroat trout, which were previously considered to belong to *Salmo*, are semelparous species in which all adults die after reproducing only once in their lifetime (semelparity). Atlantic and Pacific salmon used to exist only in the Northern Hemisphere from middle to high latitudes, but some Pacific and Atlantic salmon species were transplanted into rivers and lakes of Chile and New Zealand from the 1870s to 1900s and are now distributed in cold water regions around the world. Retropositional genome analyses indicated that pink salmon is the most advanced salmon species, while masu salmon is the most primitive species. Pink salmon are also the most widely distributed species in the northern Pacific Ocean and exhibit the largest population size, while masu salmon appear to present the most restricted distribution, in the western Pacific Ocean, and the smallest population. The relationship between salmon evolution and homing accuracy is one of the most interesting questions from a viewpoint of salmon evolution. There are only 165 species classified as migratory fish. These species have acquired many important biological functions that make their successful migration possible, and most of them are valuable fisheries resources. Recent unpredictable serious climate change affecting oceans, rivers, lake levels, thermal structure, and ocean currents and acidification (seawater and freshwater) have caused many negative and a few positive impacts on important fisheries resources, habitats, and distributions of migratory fish.

REFERENCES

Aoyama, J. (2009). Life history and evolution of migration in catadromous eel (genus *Anguilla*). *Aqua-BioScience Monographs*, *2*(1), 1–42. doi:10.5047/absm.2009.00201.0001

Artamonova, V. S., Kolmakova, O. V., Kirillova, E. A., & Makhrov, A. A. (2018). Phylogeny of salmonid fishes (Salmonoidei) based on mtDNA *COI* gene sequences (Barcoding). *Contemporary Problems of Ecology*, *11*(3), 271–285. doi:10.1134/S1995425518030022

Campbell, M. A., Lopez, J. A., Sado, T., & Miya, M. (2013). Pike and salmon as sister taxa: Detailed intraclade resolution and divergence time estimation of Esociformes + Salmoniformes based on whole mitochondrial genome sequences. *Gene*, *530*(1), 57–65. doi:10.1016/j.gene.2013.07.068 PMID:23954876

Crête-Lafrenière, A., Weir, L., & Bernatchez, L. (2012). Framing the salmonidae family phylogenetic portrait: A more complete picture from increased taxon sampling. *PLoS One*, *7*(10), e-46662. doi:10.1371/journal.pone.0046662 PMID:23071608

Gross, M. R., Coleman, R. M., & McDowall, R. M. (1988). Aquatic productivity and the evolution of diadromous fish migration. *Science*, *239*(4845), 1291–1293. doi:10.1126/science.239.4845.1291 PMID:17833216

Harden-Jones, F. R. (1968). *Fish Migration*. London: Edward Arnold.

Japan International Cooperation Agency (1991). Researches on technology transfer methods. Aquaculture in Chile. *Project-Type Technology Cooperation Activity Case Series*, *48*, 1-95. (in Japanese)

Kaeriyama, K., & Ueda, H. (1998). Life history strategy and migration pattern of juvenile sockeye (*Oncorhynchus nerka*) and chum salmon (*O. keta*) in Japan: A review. *North Pacific Anadromous Fish Commission Bulletin*, *1*, 163–171.

Kaeriyama, K., & Urabe, Y. (2018). Global warming effect for migration route of Japanese chum salmon. *North Pacific Anadromous Fish Commission Technical Report*, *11*, 91–95.

Macqueen, D. J., & Johnston, I. A. (2014). A well-constrained estimate for the timing of the salmonid whole genome duplication reveals major decoupling from species diversification. *Proceedings. Biological Sciences*, *281*(1778), 20132881. doi:10.1098/rspb.2013.2881 PMID:24452024

McDowall, R. M. (1994). The origin of New Zealand's chinook salmon, *Oncorhynchus tshawytscha*. *Marine Fisheries Review*, *56*, 1–7.

Murata, S., Takasaki, N., Saitoh, M., Tachida, H., & Okada, N. (1996). Details of retropositional genome dynamics that provide a rationale for a generic division: The distinct branching of all the Pacific salmon and trout (*Oncorhynchus*) from the Atlantic salmon and trout (*Salmo*). *Genetics*, *142*, 915–926. doi:10.1006/gcen.2001.7646 PMID:8849897

Naisbett-Jones, L. C., Putman, N. F., Stephenson, J. F., Ladak, S., & Young, K. A. (2017). A magnetic map leads juvenile European eels to the Gulf Stream. *Current Biology*, *27*(8), 1236–1240. doi:10.1016/j.cub.2017.03.015 PMID:28416118

Nakabo, T., Nakayama, K., Muto, N., & Miyazawa, M. (2011). *Oncorhynchus kawamurae* "Kunimasu", a deepwater trout, discovered in Lake Saiko, 70 years after extinction in the original habitat, Lake Tazawa, Japan. *Ichthyological Research*, *58*(2), 180–183. doi:10.100710228-011-0204-8

Ou, M., Hamilton, T. J., Eom, J., Lyall, E. M., Gallup, J., Jiang, A., ... Brauner, C. J. (2015). Responses of pink salmon to CO2-induced aquatic acidification. *Nature Climate Change*, *5*(10), 950–955. doi:10.1038/nclimate2694

Quinn, T. P. (2018). *The Behavior and Ecology of Pacific Salmon and Trout* (2nd ed.). Seattle, WA: University of Washington Press.

Quinn, T. P., Kinnison, M. T., & Uniwin, M. J. (2001). Evolution of chinook salmon (*Oncorhynchus tshawytscha*) populations in New Zealand: Pattern, rate, and process. *Genetica*, *112-113*(1), 493–513. doi:10.1023/A:1013348024063 PMID:11838785

Righton, D., Westerberg, H., Feunteun, E., Økland, F., Gargan, P., Amilhat, E., ... Aarestrup, K. (2016). Empirical observations of the spawning migration of European eels: The long and dangerous road to the Sargasso Sea. *Science Advances*, *2*(10), e1501694. doi:10.1126ciadv.1501694 PMID:27713924

Shedko, S. V., Miroshnichenko, I. L., & Nemokova, G. A. (2013). Phylogeny of salmonids (Salmoniformes: Salmonidae) and its molecular dating: Analysis of MtDNA Data. *Russian Journal of Genetics*, *49*(6), 623–637. doi:10.1134/S1022795413060112 PMID:24450195

Smith, G. R., & Stearley, R. F. (1989). The classification and scientific names of rainbow trout and cutthroat trouts. *Fisheries (Bethesda, Md.)*, *14*(1), 4–10. doi:10.1577/1548-8446(1989)014<0004:TCASNO>2.0.CO;2

Tsukamoto, K., Chow, S., Otake, T., Kurogi, H., Mochioka, N., Miller, M. J., ... Tanaka, H. (2011). Oceanic spawning ecology of freshwater eels in the western North Pacific. *Nature Communications*, *2*(1), 179. doi:10.1038/ncomms1174 PMID:21285957

Tsukamoto, K., Nakai, I., & Tesch, W. V. (1998). Do all freshwater eels migrate? *Nature*, *396*(6712), 635–636. doi:10.1038/25264

Ueda, H. (2018). Migration and Navigation in Fish. In M. K. Skinner (Ed.), *Encyclopedia of Reproduction* (2nd ed.; Vol. 6, pp. 84–89). Elsevier. doi:10.1016/B978-0-12-809633-8.20540-2

Ueda, H., & Tsukamoto, K. (Eds.). (2014). *Physiology and Ecology of Fish Migration*. Boca Raton, FL: CRC Press.

Williams, C. R., Dittman, A. H., McElhany, P., Busch, D. S., Maher, M. T., Bammler, T. K., ... Gallagher, E. P. (2019). Elevated CO_2 impairs olfactory-mediated neural and behavioral responses and gene expression in ocean-phase coho salmon (*Oncorhynchus kisutch*). *Global Change Biology*, *25*(3), 963–977. doi:10.1111/gcb.14532 PMID:30561876

Chapter 2
Hormones in Salmon

ABSTRACT

Hormones are important signaling molecules produced and secreted in the endocrine system that show interesting close similarities between humans and salmon. They are transported to target organs where they bind to their receptors and control physiological regulation and behavioral activity to maintain homeostasis via feedback mechanisms. Various hormones control freshwater and seawater adaptations to maintain water and salt balances. The juvenile imprinting migration and adult homing migration of salmon are mainly controlled by the brain (thyrotropin-releasing hormone)-pituitary (thyrotropin)-thyroid (thyroid hormones) axis and the brain (gonadotropin-releasing hormone)-pituitary (gonadotropin)-gonad (steroid hormones) axis, respectively. This chapter describes hormone species and actions, hormonal control of freshwater and seawater adaptations, and hormonal changes during juvenile imprinting migration and adult homing migration in salmon.

DOI: 10.4018/978-1-7998-2054-3.ch002

INTRODUCTION

Hormones are defined as signaling molecules that are produced and secreted by endocrine cells in endocrine glands and are transported by several circulatory systems to target cells in target organs, where they exert their functions by binding to specific receptors to control physiological regulation and behavioral activity in organisms. Hormones are the key mediators and integrators of environmental and internal information in organisms for the maintenance of "homeostasis", which is the optimal steady state of internal conditions controlling various essential functions that support life in organisms. Hormones are classified by their secretion and transportation patterns and chemical structures and show close similarities between humans and salmon. Although some endocrine glands disappeared during the evolution of animals from life under water to life on land, many endocrine glands have essential common important functions in vertebrates.

Salmon show dynamic and complex life cycles; they are born in freshwater, undergo downstream migration into seawater as juveniles for growth, during which they develop seawater adaptability, and undergo upstream migration into freshwater as adults for reproduction, during which they develop freshwater adaptability. Moreover, salmon exhibit amazing natal stream imprinting and homing abilities during downstream migration in juveniles and upstream migration in adults, respectively. These dramatic physiological and behavioral changes are mainly controlled by many hormones secreted from various endocrine glands for freshwater and seawater adaptation, natal stream imprinting during downstream migration in juveniles and natal stream homing during spawning migration in adults. Using anadromous chum salmon and lacustrine sockeye salmon as the main species, the hormonal control mechanisms of seawater and freshwater adaptability as well as the hormonal changes associated with juvenile imprinting migration during downstream migration and adult homing migration during spawning migration are explained in this chapter.

Hormone Species and Actions

The secretion and transportation patterns of hormones are classified into four types. In the endocrine type, hormones are produced and secreted from endocrine cells in the endocrine glands and transported via the blood circulatory system. In the autocrine type, hormones are produced by a cell and affect the same cell, as observed in cancer tumor cells showing out-of-control development. In the paracrine type, hormones are secreted by cells and affect nearby cells, as observed in the islets of Langerhans in the pancreas, controlling blood glucose levels via insulin and glucagon. In the neuroendocrine type, nerve cells produce hormones that affect target cells either directly or indirectly via the blood circulatory system (Figure 1).

One of the most important functions of hormones is the maintenance of "homeostasis". The principle concept of homeostasis was developed by Claude Bernard, a French physiologist from the 18th century, and Walter

Figure 1. Secretion and transportation patterns of hormones from endocrine cells to target cells showing endocrine, paracrine, autocrine, and neuroendocrine types.

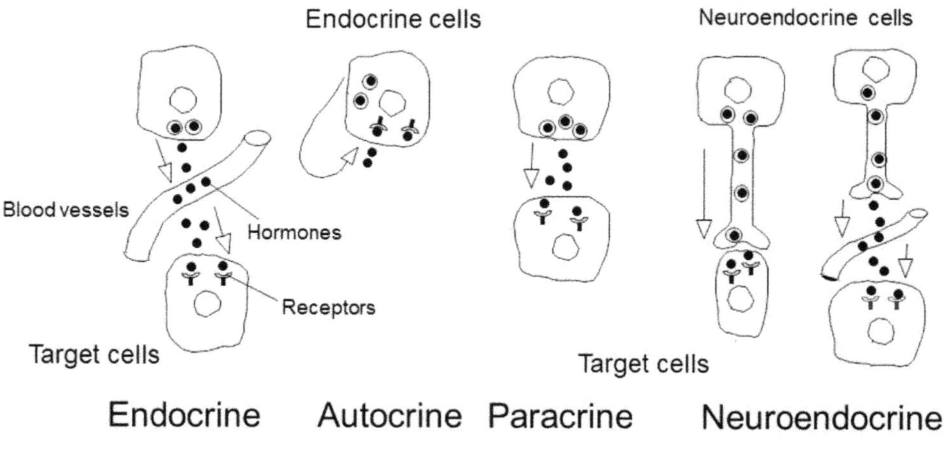

B. Cannon, an American physiologist, who established his theories in 1932 (Cannon, 1932). Homeostasis is the optimal steady state of internal conditions controlling various essential functions and systems that support life in animals, including fishes, such as body temperature, ion exchange, fluid balance, neurotransmission, and the neuroendocrine system. For the maintenance of homeostasis, endocrine systems mainly possess negative feedback mechanisms that generally promote stability. In contrast, positive feedback tends to lead to instability.

Hormones are mainly classified into five chemical structures: steroids, peptides/proteins, iodinated tyrosine derivatives, amines, and eicosanoids (Table 1). Hormones are released from various endocrine glands in all vertebrates from fish to humans. Approximately half a decade ago, Dr. Howard A. Bern, who was a great pioneer of comparative endocrinology, performed a comparison of endocrine glands between humans and trout (Bern, 1967), demonstrating close similarities between these two species. Recent achievements in comparative endocrinology in fish allow a new diagram to be drawn, showing 11 endocrine glands: brain (hypothalamus),

Table 1. Classification of hormones based on chemical structures

Chemical structures	Category	Representative hormones
Steroids	Sex hormones	Androgens: testosterone (Te), 11-ketotestosterone (11KT) Estrogenes: estradiol-17β (E2), estoriol, estrone Progestogens: progesterone, 17α,20β-dihydroxy-4-pregnen-3-one (DHP)
	Corticosteroids	Glucocorticoids: cortisol, cortisone Mineralcorticoids: aldosterone, desoxycorticosterone
	Neurosteroids	17α-hydroxypregnenolone
	Insect steroids	Ecdysteroids: ecdysterone
	Plant steroids	Alkaloids, Melanthiacease, Cardiac glycosides, Phytostroids, Brassinosteroids
	Vitamin D	Vitamin D_2 in fungi, Vitamin D_3 in animals
Peptides/Proteins	Brain (Hypothalamic) hormones	GnRH, TRH, GHRH, CRH, GnIH, somatostatin, vasotocin, oxytocin, kisspeptin
	Pituitary hormones	GTH (LH, FSH), TSH, ACTH, GH, PRL, SL; MSH
	Parathyroid hormone	PTH
	Parafollicular cells	Calcitonin
	Pancreatic hormones	Insuline, Glucagon
	Heart hormone	ANP
	Gastrointestinal hormones	Gastrin, Secretin
	Stannius corpuscle hormone	Stanniocalcin
Iodinated tyrosine derivatives	Thyroid hormone	T4, T3
Amines	Catecholamine	Adrenaline, Noreadrenaline, Dopamin, Epinephrine, Norepinephrine
	Tryptamine	Melatonine, Serotonine
	Imidazole	Histamine
	Choline	Acetylcholine
Eicosanoids	Prostaglandin (PG)	PGI_2, PGD_2, PGE_2, $PGF_{2\alpha}$
	Thromboxane	Thromboxane
	Leukotriene (LT)	Cysteinlyl leukotrienes (LTC_4, LTD_4, LTF_4), LTB_4, LTB_5, LTG_4

Table 2. Classification of endocrine gland, hormones and their main functions in salmon

Endocrine gland	Hormone	Main functions
Brain (Hypothalamus)	Releasing hormone (RH)	Control the release of pituitary hormones
	Inhibiting hormone (IH)	Control the inhibition of pituitary hormones
	Vasotocin	Control water balance and osmoregulation
	Isotocin	Control water balance and osmoregulation
	Kisspeptin	Control the release of GnRH
Pituitary gland	ACTH	Producrtion and release of cortisol from interrenal gland
	GH	Regulation of growth directly or through IGF-1
	PRL	Control the osmoregulation to adapt fresh water
	SL	Calcium/phosphate mobilization and energy homeostasis
	GTH	FSH: Stimulation of vitellogenesis and spermatogenesis
		LH: Stimulation of final gonadal maturation
	TSH	Production and release of T3 and T4 in the thyroid gland
	MSH	Production and release of melatonin in the melanocytes
Pineal gland	Melatonin	Regulation of circadian rhythm
Thyroid gland	T3, T4	Regulatontion of metabolism and control of metamorphosis
Parathyroid gland	PTH	Regulation of serum calcium and phosphate
Ultimobranchial glnd	Calcitonin	Regulation of serum calcium and phosphate
Heart	ANP	Reduction in extracellular fluid volum
Gastrointestinal tract	Gastrin, Secretin	Stimulation to secrete gastric acid, Regulation of water homeostasis
Pancreatic islets	Insuline, Glucagon	Metabolism of glucose
Interrenal gland	Aldosterone, Cortisol	Stress response
Chromaffin cells	Catecholaminses	Control the sympathetic nervous system response
Gonads	Steroid hormones	Control of sex channge, gonadal development and maturation
Stannium Corpuscles	Stanniocalcin	Regulation of serum calcium and phosphate levels
Caudal neurosecretory system	Ulotensin	Regulation of water/ion balance

pituitary gland, pineal gland, thyroid gland, ultimobranchial gland (parafollicular cells), heart, gastrointestinal tract, pancreatic islets (Islets of Langerhans), adrenal cortex (interrenal gland), adrenal medulla (chromaffin cells), and gonads. These glands commonly control many similar endocrine functions in humans and salmons (Figure 2). Fish do not have a parathyroid gland, but several fish species have been found to express parathyroid hormone (PTH) in their gills. The Stannius corpuscles and caudal neurosecretory system are limited to water-living fish. The hormones secreted by various endocrine glands and their main functions in salmon are summarized in Table 2.

The autonomic nervous system (ANS) consists of the sympathetic nervous system and the parasympathetic nervous system, which are characterized by the neurotransmitter, adrenaline/noradrenaline and acetylcholine, respectively. The ANS has important functions in the maintenance of homeostasis in collaboration with the endocrine system.

Figure 2. Comparison of endocrine glands of human beings and salmon as modified from Bern (1967) adding new relationships of endocrine systems between human beings and salmon.

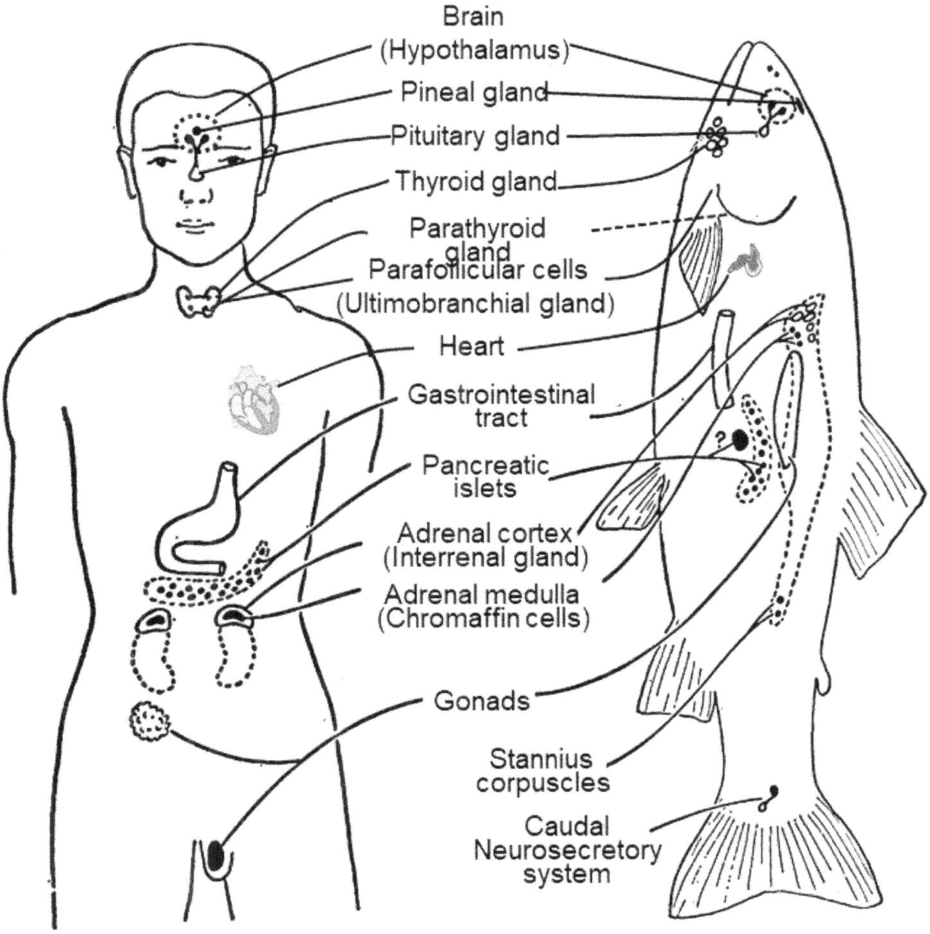

Mechanisms of Freshwater and Seawater Adaptation

There are is a wide range of ion concentrations and osmotic pressures between freshwater and seawater in which euryhaline teleost fishes can survive (Kaneko et al., 2008). In general, freshwater teleosts face water loading and salt loss through their permeable body surfaces, and they discharge excess water by producing a large volume of hypotonic urine

Figure 3. Osmoregulation of freshwater and seawater teleosts fishes showing movement of water and salts in freshwater and seawater teleosts, which produced by Prof. Toyoji Kaneko.

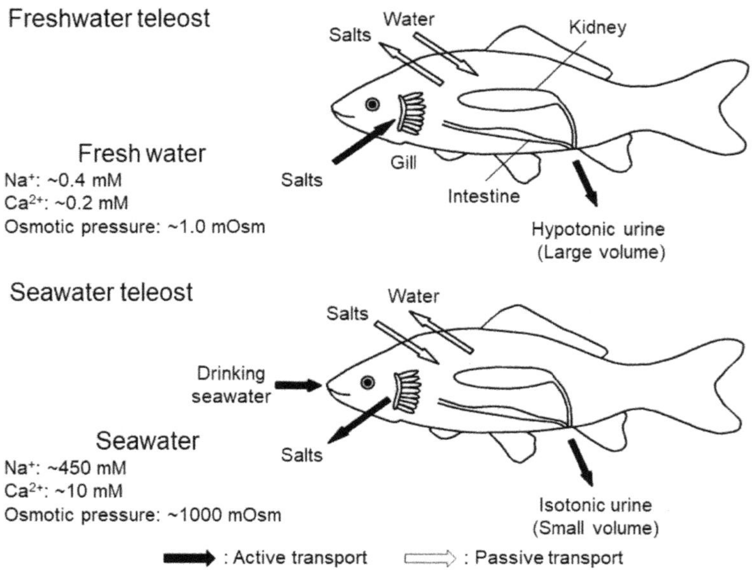

in the kidney and absorb salts thorough the gill epithelia. In contrast, seawater teleosts face water loss and salt loading, and they drink seawater and absorb water in the intestine by producing a small volume of isotonic urine, while the excess salts are actively excreted from the gill epithelia (Figure 3). However, the osmotic pressure of freshwater and seawater teleosts is controlled by various hormones for maintenance at approximately 300 mOsm-H_2O. For freshwater adaptation in teleosts, prolactin (PRL) secreted from the pituitary gland mainly regulates water and salt balance (Hirano, 1986; Sanyal & Sen, 2018). In freshwater salmon, PRL controls freshwater adaptability by preventing salt loss in the chloride cells of the gills, increasing the glomerular filtration rate in the kidney, increasing salt excretion and decreasing water retention in the tubule and urinary bladder, and producing a large volume of hypotonic urine. In contrast, cortisol secreted from the interrenal gland and growth hormone (GH) secreted from the pituitary gland control seawater adaptation in teleosts. In seawater salmon, cortisol controls the differentiation of chloride cells in the gills, the activation of Na^+/K^+-ATPase (an enzyme in the plasma membrane that pumps sodium out of cells and potassium into cells) and the promotion of salt excretion in the chloride cells and promote water retention in the intestine. GH affects insulin-like growth factor I (IGF-I) produced in the liver, which promotes the activation of Na^+/K^+-ATP ase in the chloride cells (McCormick, 2009).

Interestingly, PRL and GH evolved through the duplication of an ancestral gene approximately 400-800 million years ago (Power, 2005), and PRL and GH in teleosts play important roles in freshwater and seawater adaptation, respectively. During vertebrate development, PRL evolved to stimulate milk production in the mammary glands of female mammals. The functions of PRL in freshwater adaptation in salmon and lactation in mammals are apparently unrelated; however, the common functions of PRL include the determination of the permeability of water and substances (salt and protein) in the cells of many organs, where regulation by PRL prevents substance outflow and stimulates water outflow. In vertebrates, GH presents a common function related to IGF-I production to stimulate cell differentiation, cell growth and development, and electrolyte metabolism. Specifically, salmon grow faster in seawater than freshwater. Cortisol is known as a stress hormone in humans that increases blood sugar levels and blood pressure, suppresses the immune system and aids in the metabolism of fat, protein, and carbohydrates. Juvenile salmon encounter environmental stress due to the large differences in ion concentrations and osmotic pressures between freshwater and seawater during downstream migration, when cortisol can act as the stress hormone.

Hormonal Changes in Juvenile Salmon During Imprinting Migration

Natal stream imprinting during downstream migration in juvenile salmon is associated with the PST (smoltification), which is a complex physiological change that results in seawater adaptability. Smoltification is considered a kind of metamorphosis, as tadpoles change into toads and is regulated by thyroxine (T4) produced in the thyroid glands. However, if juvenile salmon cannot undergo downstream migration to the sea for some reason, they should return to the parr form and lose their seawater adaptability

(Björnsson et al., 2012). Smoltification is regulated by endocrine hormones such as growth hormone-releasing hormone (GHRH), corticotropin-releasing hormone (CRH), and thyrotropin-releasing hormone (TRH) in the brain; GH, adrenocorticotropic hormone (ACTH), and thyroid-stimulating hormone (TSH) in the pituitary gland; and IGF-I, cortisol, and thyroid hormones (T4 and triiodothyronine: T3) in many of the endocrine organs (Björnsson et al., 2011). Specifically, the brain-pituitary-thyroid (BPT) axis is thought to play critical roles during the imprinting process that occurs during downstream migration. TRH in the brain stimulates TSH release in the pituitary gland, which in turn stimulates thyroid hormone (T4 and T3) release in the thyroid gland (Figure 4A). There is a feedback mechanism in which thyroid hormone controls TRH and TSH activity to maintain homeostasis of the BPT axis.

Figure 4. The brain-pituitary-thyroid (BPT) axis (A) and brain-pituitary-gonad (BPG) axis (B) in Pacific salmon. Thyrotropin-releasing hormone (TRH) in the brain stimulates thyrotropin (TSH) release in the pituitary gland that stimulate thyroid hormone (T4 and T3) release in the thyroid gland. Gonadotropin-releasing hormone (GnRH) stimulate gonadotropin (GTH) release in the pituitary gland that stimulate steroid hormone (E2, Te, 11KT, and DHP) release in the gonad. Thyroid hormone controls TRH and TSH activities in the BPT axis, and steroid hormone controls GnRH and GTH activities in the BPG axis by feedback mechanisms.

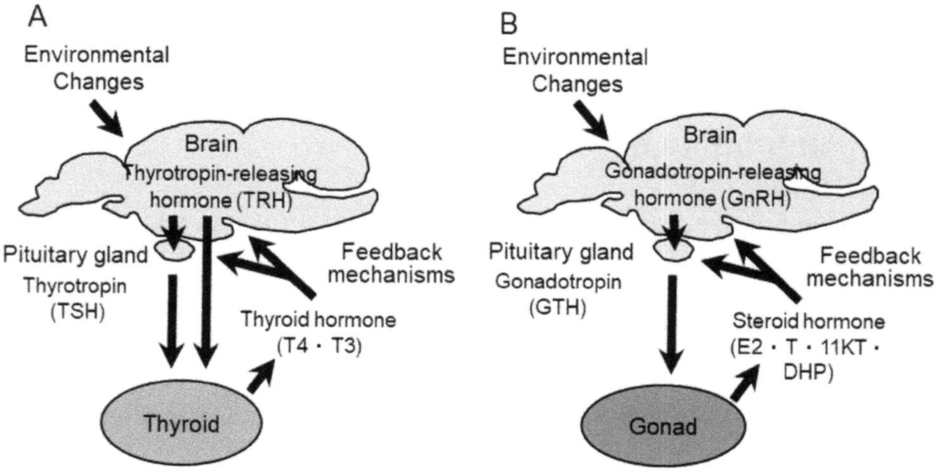

Several studies have described the hormone profiles that occur during downstream migration as well as the artificial induction of downstream movement in juvenile Pacific salmon. In coho salmon in California, a distinct surge in plasma T4 during smoltification has been reported to be coincident with the new moon (Grau et al., 1981). A similar serum T4 surge was detected in smolts of masu salmon in Hokkaido (Yamauchi et al, 1985). A clear T4 surge was observed during the downstream migration of chum salmon fry, suggesting that T4 plays an important role in the preparation for downstream migration. However, T4 has not been identified as the sole factor responsible for the onset of downstream migration (Iwata et al., 2003; Ojima & Iwata, 2007). Ojima & Iwata (2009) reported that the central administration of GHRH, CRH, melatonin, N-acetyl serotonin, and serotonin stimulated downstream movements, while only GHRH stimulated schooling behaviors. In juvenile Chinook salmon, the central administration of CRH was also found to induce downstream movements (Clements & Schreck, 2004). Additionally, treatments that affect thyroid hormones have been shown to prompt olfactory cellular proliferation in coho salmon (Lema & Nevitt, 2004). In fall-smolting coho salmon, the central administration of GHRH and CRH stimulated downstream movement and T4 secretion (Ojima & Iwata, 2010). In the same species, the levels of *tshβ* mRNA and TSH in the pituitary gland and the levels of TSH, T4 and T3 in the plasma have been assessed and were found to be correlated with smoltification (Larsen et al., 2011). In the brains of sockeye salmon, the mRNA that encodes the *trh* precursor has been studied via *in situ* hybridization (Ando et al., 1998), and the expression of the cDNAs encoding TRH receptor subtypes has been examined (Saito et al., 2011). T3 has been reported to determine the onset of the sensitive period for imprinting, and T3 plays a critical role in later learning in chicks (Yamaguchi et al., 2012). Although the onset of downstream migration, which includes smoltification, is regulated by a complex set of several endocrine hormones, BPT hormones are considered to play a critical role in imprinting during downstream migration (Dittman & Quinn, 1996).

The hormonal profiles in the BPT axis in underyearling juvenile chum salmon were investigated from the Chitose Hatchery (Site **A**) to Ishikari Bay (Site **F**) (Figure 5). Changes in the BPT hormone profiles were measured based on the whole-brain mRNA expression of *trha/b* and the *tshβ* in the upper head, which contains the pituitary gland, and the levels of T4 and T3 isolated from the lower jaw, which contains the thyroid glands. Ueda et al. (2016) revealed that the environmental changes experienced

Figure 5. Map showing the sampling sites of juvenile chum salmon during downstream migration and adult chum salmon during homing migration. A: Chitose Hatchery; B: second bridge of the Chitose; C: Chitose River in Kamaka Ward, Chitose; D: Confluence point of the Chitose and Old Yubari Rivers; E: Mouth of the Ishikari River; F: Ishikari Bay; G: Bering Sea; H: coastal area of Atsuta; I: confluence point of the Ishikari and Chitose Rivers. J: Indian waterwheel.

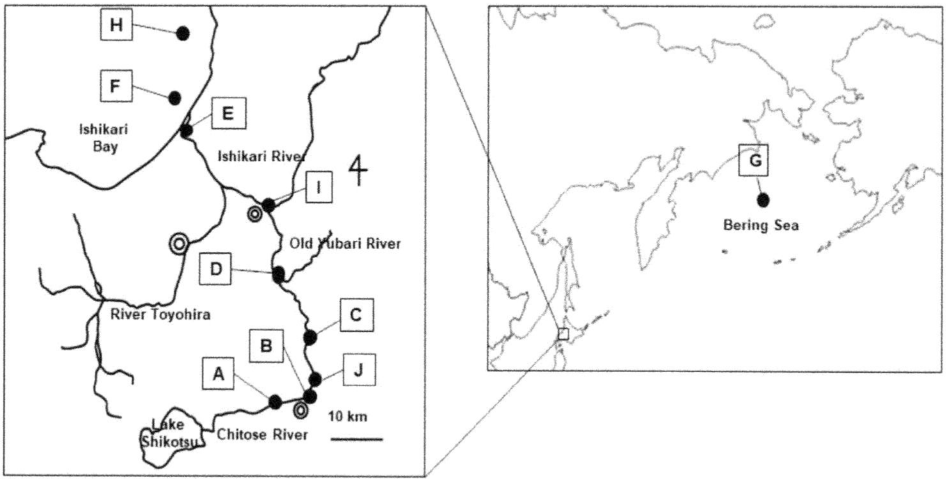

by fish during release from the hatchery to the natal stream significantly increased *trha/b* expression levels and that *tshβ* expression levels and T4/T3 levels significantly increased during downstream migration towards the sea (Figure 6).

Trha gene expression increased gradually during rearing in the hatchery and peaked significantly immediately after release into the Chitose River at Site **B**. In contrast, *trhb* gene expression showed no changes in the hatchery but presented a significant increase after release at Site **B**. Although the precise roles of TRHa and TRHb are unknown, TRHb appears to react to acute environmental changes, such as release to a river. *tshβ* gene expression showed a gradual increase at Site **B** and a significant peak at Site **E**. T4 and T3 levels then increased significantly at Site **F**. These data support the importance of the BPT axis in the early downstream migratory period of chum salmon and are consistent with the hypothesized role of thyroid hormones during the imprinting phase in chum salmon.

Figure 6. Changes in the profiles of the brain-pituitary-thyroid hormones in juvenile chum salmon during downstream migration. Fish were sampled at the Chitose Hatchery (Site A) at four time points (January (A1), February (A2), March (A3), and April (A4)) and following release into the Chitose River from Site B to Site F. Data are presented as the means ± SEM (N=10-15). Different letters represent significant differences at p<0.05 level.

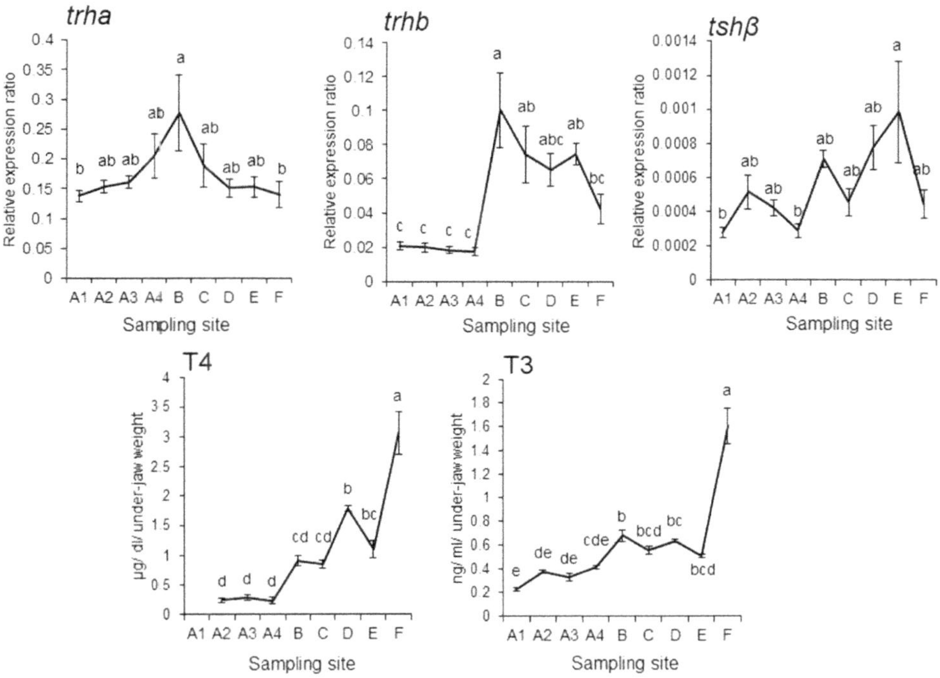

Hormonal Changes in Adult Salmon During Homing Migration

Homing migration from the feeding habitat to the breeding habitat in the natal stream in adult salmon is closely related to gonadal maturation and is mainly regulated by the brain-pituitary-gonad (BPG) axis (Figure 4B). Two gonadotropin-releasing hormones (GnRH), i.e., salmon GnRH (sGnRH; commonly named GnRH3) and chicken GnRH-II (cGnRH-II; commonly named GnRH2), exist in various regions of the brains of salmon (Amano et al., 1997). Specifically, sGnRH neurons, which exhibit two different precursor genes (i.e., sGnRH-I and sGnRH-II) are found in the olfactory bulb (OB), the terminal nerve (TN), and the preoptic area (POA), and these neurons are considered to play different roles in salmon homing migration; however, their exact roles have not yet been clarified. sGnRH in the POA is believed to mainly control the synthesis and release of gonadotropic hormone (GTH), luteinizing hormone (LH) and follicle-stimulating hormone (FSH). GTHs induce steroidogenesis in the gonads, and steroid hormones stimulate gametogenesis and final gameto-maturation; furthermore, estradiol-17β (E2) and testosterone (Te) are active during vitellogenesis; Te and 11-ketotestosterone (11KT) are active during spermatogenesis, and 17α,20β-dihydroxy-4-pregnen-3-one (DHP) is active during final gameto-maturation in both sexes (Nagahama, 1997). There is a feedback mechanism whereby steroid hormones control GnRH and GTH activities to maintain the homeostasis of the BPG axis.

Anadromous Adult Chum Salmon from the Bering Sea to the Chitose River, Hokkaido, Japan

The hormone profiles in the BPG axis of adult chum salmon migrating from the Bering Sea (Site **G**) to the natal hatchery in the Chitose River (Site **A**) (Figure 5) were measured using specific time-resolved fluoroimmunoassays ((TR-FIA) (Yamada et al., 2002). The level of sGnRH in the OB of both sexes peaked when the fish were located between Sites **E** and **F**, where the olfactory discrimination of the natal stream is expected to be important. Peak sGnRH in the telencephalon (TE) was observed when the fish were at the branch point of the Chitose River from the Ishikari River (Site **I**), where olfactory functions should be highly activated as individuals attempt to determine which branch to enter (Figure 7). In the pituitary gland, sGnRH levels tended to increase in concert with LH levels around Site **F** and **E**. In contrast, FSH levels did not show any clear correlations with sGnRH levels in the pituitary gland. Although the roles of cGnRH-II in these brain regions are currently unclear, their levels increased in the medulla oblongata (MO) in both sexes at Site **I** and **J**, and their levels in the optic tectum (OT) increased in males at Site **J**. In the diencephalon (DC) and cerebellum (CB), cGnRH-II levels showed no significant changes during homing migration (Figure 7) (Ueda, 2011).

Figure 7. Changes in salmon gonadotropin-releasing hormone (sGnRH) contents in the olfactory bulb (OB) and telencephalon (TE) (A); sGnRH, luteinizing hormone (LH), and follicle-stimulating hormone (FSH) in the pituitary gland (B); chicken GnRH-II (cGnRH-II) in the optic tectum (OT), the diencephalon (DC), the cerebellum (CB), and the medulla oblongata (MO) (C) of adult male and female chum salmon during homing migration from Site G to Site A. Data are presented as the means ± SEM (N=8-12).

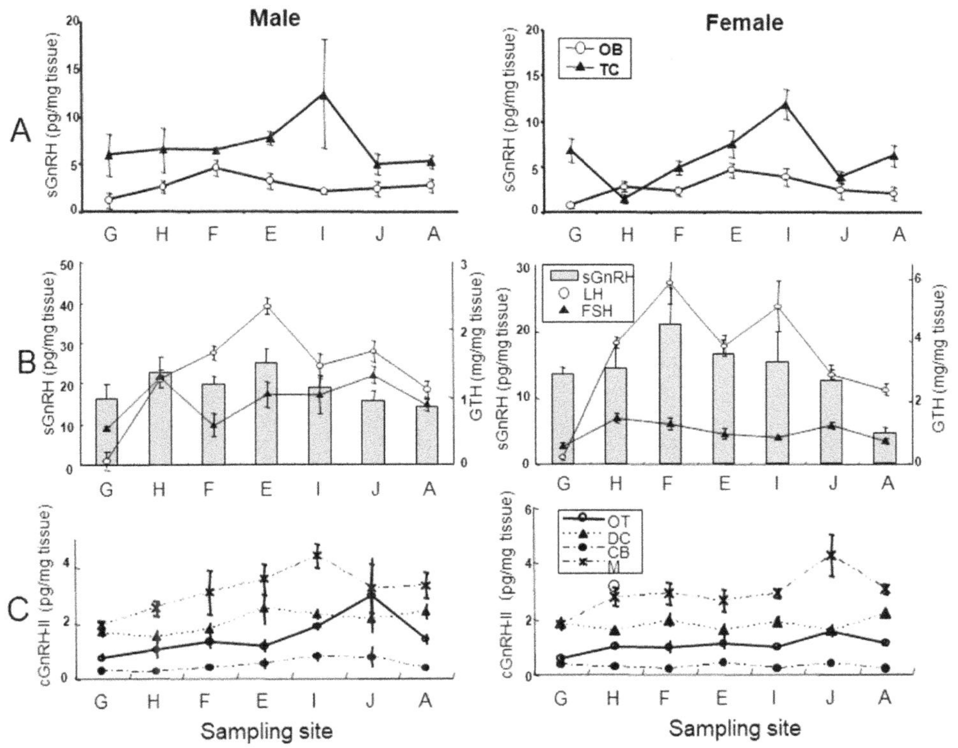

sGnRH neurons showing signals of pro-*sgnrh* mRNA were observed in the olfactory nerve (ON) of chum salmon at Site **F** but not in fish at Site **A**, and the intensity of the signals was high at Site F. In contrast, the numbers and intensity of the signals of sGnRH neurons in the TE and POA increased at Site A (Figure 8). These different sGnRH neuron numbers and intensities suggest different roles of sGnRH during homing migration in adult chum salmon (Kudo et al., 1996). Changes in the levels of *gth* subunit mRNAs in the pituitary gland of prespawning chum salmon

*Figure 8. Schematic illustrations of the distribution of sGnRH neurons and changes in the number of neurons showing signals for number of sGnRH neurons and intensity of pro-sGnRH signals in chum salmon at Site F and Site A. Data are presented as the means ± SEM (N=3-7). Significant differences are indicated at *p<0.05 and **p<0.01 levels. ND: not detected; OB: olfactory bulb; OB-TE: transitional area between olfactory bulb and telencephalon; ON: olfactory nerve; ON-OB: transitional area between olfactory nerve and olfactory bulb; POA: preoptic area; TE: telencephalon.*

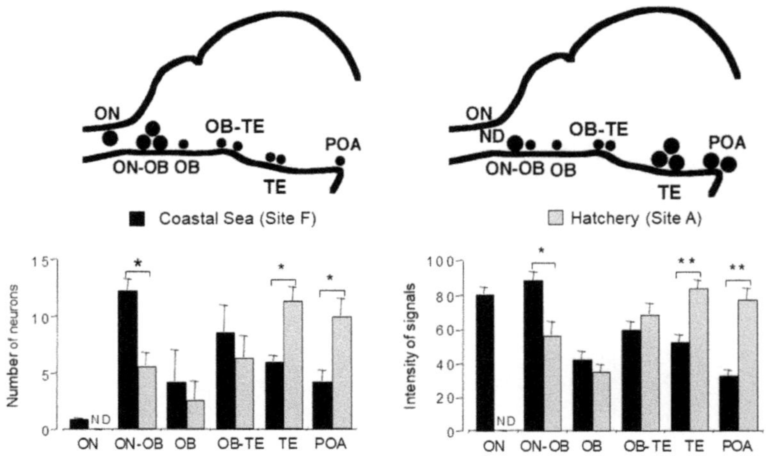

demonstrated that the levels of *gthα2* (there are two distinct α subunit proteins, and only α2 subunit is utilized as the composer of LHβ) and *lhβ* increase when an individual enters freshwater, but no changes in the *fshβ* level were observed (Kitahashi et al. 1998a). Interestingly, sGnRH levels in the TC and serum Te levels in both sexes showed a coincident peak at the branch point of the Chitose River from the Ishikari River (Site **I**). Ueda et al. (2016) showed that the mRNA expression levels of *sgnhr-I* and *sgnrh-II* in the OB and TE appeared to increase during homing migration from the Bering Sea (Site **G**) to the natal hatchery in the Chitose River (Site **A**), with some differences between males and females that might be related to sex-specific gonadal maturation (Figure 9). Interestingly, *sgnrh-II* gene expression in the hypothalamus of males was also high in the Bering Sea (Site **G**). These results suggest that sGnRH secreted in different brain regions is involved in brain region-dependent functions during homing migration in adult chum salmon: sGnRH neurons in the

Figure 9. Changes in the gene expression profiles of salmon gonadotropin-releasing hormone (sGnRH)-I and -II in the olfactory bulb and telencephalon of adult chum salmon during homing migration from Site G to Site A. Data are presented as the means ± SEM (N=10). Different letters represent significant differences at p<0.05 level.

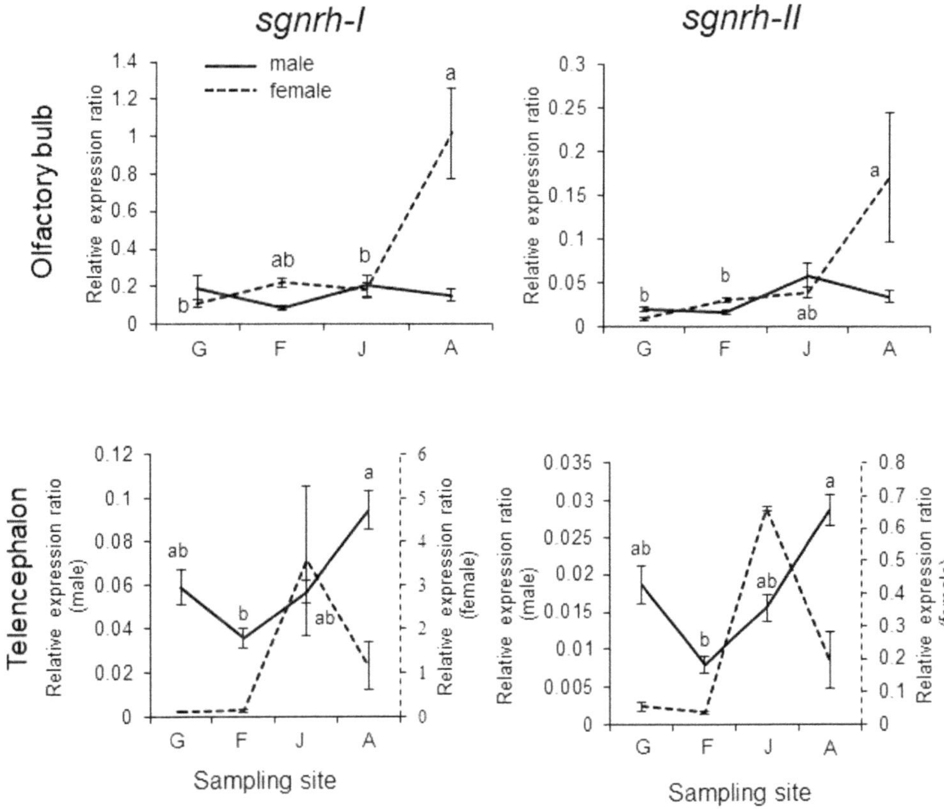

ON and OB may exhibit critical functions in the olfactory discrimination of the natal stream, while those in the TE and POA may exhibit critical functions in gonadal development and final maturation.

Serum steroid hormone levels showed similar profiles to those observed in our previous studies (Ueda et al. 1984, Ueda 1999): E2 in females and 11KT in males increased during vitellogenesis and spermatogenesis, respectively, during homing migration from the Bering Sea to the Chitose River. Although E2 in males and 11KT in females showed interesting small peaks during upstream migration to the natal hatchery, the precise roles

Figure 10. Changes in serum steroid hormone levels of adult male and female chum salmon during homing migration from Site G to Site A. Data are presented as the means ± SEM (N=8-12). DHP: 17α,20β-dihydroxy-4-pregnen-3-one; E2: Estradiol-17β, 11KT: 11-ketotestosterone; Te: testosterone.

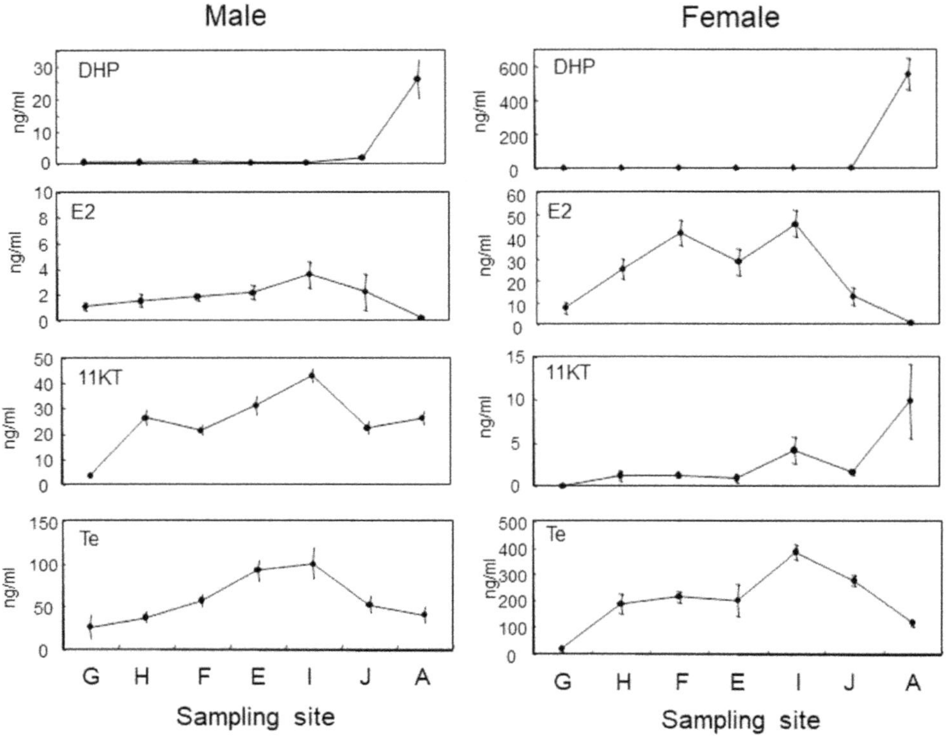

of these peaks were unknown. Te, which is a precursor of both E2 and 11KT, showed similar peaks to E2 and 11KT. DHP increased dramatically at the time of final gonadal maturation in both sexes (Figure 10). In addition, year-to-year differences in the plasma levels of steroid hormones in prespawning chum salmon were also studied in comparison with the sea surface temperature (SST) of coastal sea, and it was found that there were year-to-year differences in plasma levels of steroid hormones and gonadal maturity, some of which may be influenced by the year-to-year variation of SST (Onuma et al., 2003).

Haraguchi et al. (2015) showed that 7α-hydroxypregnenolone (7α-OH PREG), a newly identified neuronal modulator of locomotion, acts as a key factor in the upstream migration of salmon. First, 7α-OH PREG and cytochrome P450 7α-hydroxylase, a steroidogenic enzyme producing 7α-OH PREG, were identified in the salmon brain, and it was then found that 7α-OH PREG synthesis in the brain increases during upstream migration. Subsequently, it was demonstrated that 7α-OH PREG increases the upstream migratory behavior of salmon. We further found that 7α-OH PREG acts on dopamine neurons in the magnocellular preoptic nucleus during upstream migration. Thus, 7α-OH PREG stimulates upstream migratory behavior through the dopaminergic system in chum salmon. These findings provide new insights into the molecular mechanisms of upstream fish migration.

Lacustrine Adult Sockeye Salmon in Lake Shikotsu, Hokkaido, Japan

Since it is difficult to carry out experimental treatments to manipulate endocrinological functions in anadromous salmon owing to the salinity differences between seawater and freshwater, lacustrine salmon populations offer a model system for studying the hormonal control mechanisms of salmon homing. In Lake Shikotsu (surface area 78 km^2, average and maximum depths of 265 m and 363 m, respectively), Hokkaido, Japan, lacustrine sockeye salmon are artificially produced via hatchery propagation. Adult sockeye salmon were captured between September and November adjacent to their natal hatchery prior to spawning and were sampled for serum steroid hormones, tagged, and released in the center of the lake. The fish were resampled at recaptured to characterize the changes in steroid hormone levels in individual migrants and to monitor homing duration and precision in each month (Sato et al., 1997). Homing duration was significantly truncated from September to October in males and from October to November in females (Figure 11A). All males returned more rapidly than the females in September and October, although half of the males did not return to the natal site in November. In contrast, 78-90% of females returned over the entire three-month sampling period (Figure 11B). Interestingly, the average homing percentage of both sexes over three months was 83%, with no sex difference in the total number of homing individuals. Although male salmon do not display territorial behavior, they are highly aggressive to compete for access to females, suggesting that early-returning males accrue some benefits in securing females for breeding. It has been reported that increased Te levels are correlated with compromised immune function (Harris & Bird, 2000), suggesting a tradeoff in which faster migration may be advantageous even with the risk of disease.

GnRH analog (GnRHa) treatment has been reported to be highly effective in inducing GTH release, ovulation and spermiation in teleost fishes (Zohar, 1996; Zohar & Mylonas, 2001). To investigate the effect of GnRHa treatment on homing duration in lacustrine sockeye salmon in Lake Shikotsu, Sato et al. (1997) and Kitahashi et al. (1998c) investigated the effect of intramuscular GnRHa implantation on both the homing profiles and serum steroid hormone levels of fish in September. Although gonadectomy was not performed in these experiments, GnRHa implantation was highly efficient in shortening the homing duration and caused dramatic increases in serum DHP levels in both sexes. An interesting discrepancy was observed between rapidly and slowly returning males: rapidly returning males exhibited higher serum Te levels and lower serum DHP levels than slowly returning males. To examine the direct effect of T and DHP on homing duration, T and DHP were implanted in fish in September. Te implantation tended to reduce homing duration in both males and females, but no significant difference was observed. DHP implantation significantly shortened homing duration in females, but it did not have any significant effect in males (Figure 11C). Interestingly, the direct actions of Te and DHP on homing migration are sex dependent. The peak of plasma Te levels in lacustrine sockeye salmon of both sexes was observed at the time when they gathered at the mouth of their natal stream in Lake Chuzenji, Japan (Ikuta, 1996). Androgens are known to be involved in the stimulation of aggressive behavior in teleost fishes (Villars, 1983), and serum Te and 11KT are the two major androgens that influence spawning behaviors, downstream and upstream migration, and the social dominance hierarchy (Kindler et al., 1989; Cardwell & Liley, 1991; Pankhurst & Barnett, 1993; Brantly et al., 1993; Cardwell et al., 1996; Munakata et al., 2001a, b). Munakata (2012) reviewed the correlation between migratory behaviors and steroid hormones in masu salmon.

GnRHa implantation was also highly effective in accelerating gonadal maturation in anadromous, maturing sockeye salmon of both sexes. The expression of GTH subunit genes in the pituitary gland was examined and revealed that the levels of *gthα* and *lhβ* mRNAs in GnRHa-implanted

*Figure 11. Changes in homing duration (A) and percentage (B) of lacustrine sockeye salmon in Lake Shikotsu from September to November. Effects of GnRH analog (GnRHa), testosterone (Te), and 17α,20β-dihydroxy-4-pregnen-3-one (DHP) implantation on homing duration of lacustrine sockeye salmon in Lake Shikotsu in September (C). Data are presented as the means ± SEM (A and B, N=9-12; C, N=10; D, N=10). Significant differences are indicated at *p<0.05 and **p<0.01 levels.*

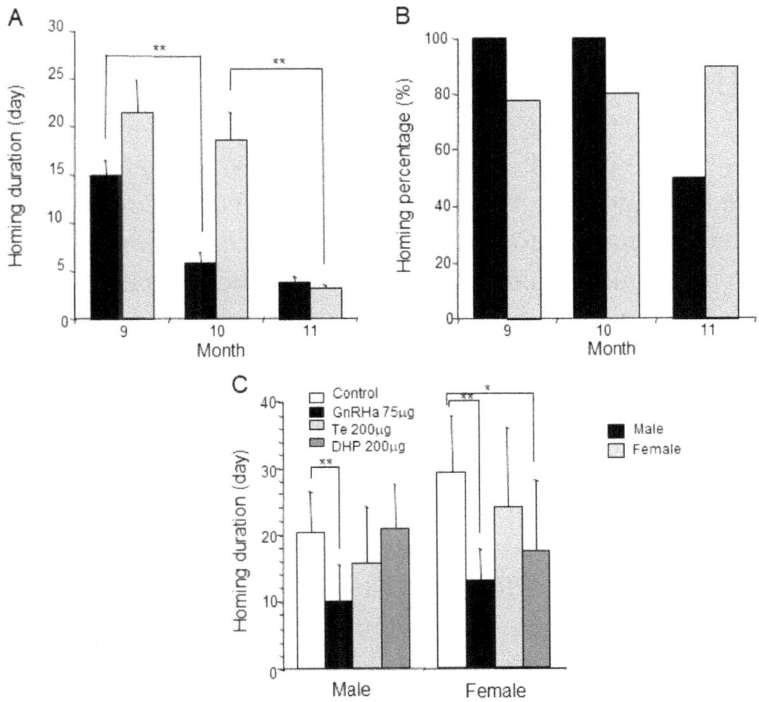

fish were higher than those in control fish, but the levels of *fshβ* mRNA showed no change (Kitahashi et al., 1998b). The implantation of GnRHa caused a significant elevation of serum DHP levels in both sexes but had no effect on the levels of Te and 11KT in males or E_2 and Te in females (Fukaya et al., 1998). These data clarify that sGnRH in the brain stimulates LH release from the pituitary gland, after which LH increases serum DHP levels in both sexes during the final phase of gamete maturation as well as the later part of homing migration. GnRH is believed to play a prominent role in the homing migration of both sexes, but gonadal steroids, especially T and DHP, seem to differentially affect the homing migration of males and females. Although DHP is known to be a maturation-inducing steroid in salmonids (Nagahama and Adachi 1985), its function in the central nervous system has not yet been clarified. The functional roles of Te and DHP in salmon homing migration should be further investigated with special attention to their action on the sensory and central nervous systems (Ueda et al., 2007).

CONCLUSION

Hormones are important signaling molecules that are produced and secreted in the endocrine system, transported to target organs where they bind to their receptors, control physiological regulation and behavioral activity to maintain homeostasis by feedback mechanisms, show several secretion and transportation patterns and are classified into five types of chemical structures: steroids, peptides/proteins, iodinated tyrosine derivatives, amines, and eicosanoids. They show interesting close similarities between humans and salmon, except in the Stannius corpuscles and caudal neurosecretory system of salmon, which disappeared during the evolution of animals from life under water to life on land. In salmon, freshwater and seawater adaptations to maintain water and salt balances are mainly controlled by PRL and GH/IGF-I, respectively, through body surfaces and gill epithelia and in the kidney. Natal stream imprinting during downstream migration is associated with smoltification, which is a complex physiological change that results in seawater adaptability and is regulated by several hormones, especially by the BPT axis. During downstream migration from the Chitose Hatchery to Ishikari Bay in juvenile chum salmon, the hormone profiles of TRH in the brain, TSH in the pituitary gland, and T3/T4 in the thyroid gland show dramatic changes. Natal stream homing during spawning and upstream migration is associated with gonadal maturation, which is mainly regulated by the BPG axis. During homing migration from the Bering Sea to the Chitose Hatchery in anadromous adult chum salmon, sGnRH secreted in different brain regions is involved in important brain region-dependent functions. In the pituitary gland, sGnRH levels tended to increase in concert with LH levels during upstream migration in the Ishikari and Chitose Rivers, but FSH levels did not show any clear correlations with sGnRH levels. Serum E2 in females and 11KT in males increased during vitellogenesis and spermatogenesis, respectively, during homing migration from the Bering Sea to the Chitose River, and DHP increased dramatically at the time of final gonadal maturation in both sexes. A newly identified neuronal modulator, 7α-OH PREG, stimulates upstream migratory behavior through the dopaminergic system in chum salmon. In lacustrine sockeye salmon in Lake Shikotsu, GnRHa implantation was highly efficient in shortening the homing duration and caused dramatic increases in serum Te and DHP levels in a sexually differentiated manner. The critical roles of the hormonal control mechanisms of the BPT and BPG axes in imprinting and homing migration, respectively, in salmon should be examined in future physiological and molecular biological studies.

REFERENCES

Amano, M., Urano, A., & Aida, K. (1997). Distribution and function of gonadotropin-releasing hormone (GnRH) in the teleost brain. *Zoological Science*, *14*(1), 1–11. doi:10.2108/zsj.14.1 PubMed

Ando, H., Ando, J., & Urano, A. (1998). Localization of mRNA encoding thyrotropin-releasing hormone precursor in the brain of sockeye salmon. *Zoological Science*, *15*(6), 945–953. doi:10.2108/zsj.15.945

Bern, H. A. (1967). Hormones and endocrine glands of fishes. *Science*, *158*(3800), 455–462. doi:10.1126/science.158.3800.455 PubMed

Björnsson, B. T., Einarsdottir, I. E., & Power, D. (2012). Is salmon smoltification an example of vertebrate metamorphosis? Lessons learnt from work on flatfish larval development. *Aquaculture (Amsterdam, Netherlands)*, *362-363*, 264–272. doi:10.1016/j.aquaculture.2011.03.002

Björnsson, B. T., Stefansson, S. O., & McCormick, S. D. (2011). Environmental endocrinology of salmon smoltification. *General and Comparative Endocrinology*, *170*(2), 290–298. doi:10.1016/j.ygcen.2010.07.003 PubMed

Brantly, R. K., Wingfield, J. C., & Bass, A. H. (1993). Sex steroid levels in Porichthys notatus, a fish with alternative reproductive tactics, and a review of the hormonal bases for male dimorphism among teleost fishes. *Hormones and Behavior*, *27*(3), 332–347. doi:10.1006/hbeh.1993.1025 PubMed

Cannon, W. B. (1932). The Wisdom of the Body. New York: W.W. Norton & Company; doi:10.1097/00000441-193212000-00028.

Cardwell, J. R., & Liley, N. R. (1991). Androgen control of social status in males of a wild population of stoplight parrotfish, Sparisome viride (Scaridae). *Hormones and Behavior*, *25*(1), 1–18. doi:10.1016/0018-506X(91)90035-G PubMed

Cardwell, J. R., Sorensen, P. W., Van Der Kraak, G. J., & Liley, N. R. (1996). Effect of dominance status on sex hormone levels in laboratory and wild-spawning male trout. *General and Comparative Endocrinology*, *101*(3), 333–341. doi:10.1006/gcen.1996.0036 PubMed

Clements, S., & Schreck, C. B. (2004). Central administration of corticotropin-releasing hormone alters downstream movement in an artificial stream in juvenile Chinook salmon (Oncorhynchus tshawytscha). *General and Comparative Endocrinology*, *137*(1), 1–8. doi:10.1016/j.ygcen.2004.02.004 PubMed

Dittman, A. W., & Quinn, T. P. (1996). Homing in Pacific salmon: Mechanisms and ecological basis. *The Journal of Experimental Biology*, *199*, 83–91. Retrieved from jeb.biologists.org/content/199/1/83 PubMed

Fukaya, M., Ueda, H., Sato, A., Kaeriyama, M., Ando, H., Zohar, Y., ... Yamauchi, K. (1998). Acceleration of gonadal maturation in anadromous maturing sockeye salmon by gonadotropin-releasing hormone analog implantation. *Fisheries Science*, *64*(6), 948–951. doi:10.2331/fishsci.64.948

Grau, E. G., Dickhoff, W. W., Bern, H. A., & Folmar, L. C. (1981). Lunar phasing of the thyroxine preparatory to seaward migration of salmonid fish. *Science*, *211*(4482), 607–609. doi:10.1126/science.7455703 PubMed

Haraguchi, S., Yamamoto, Y., Suzuki, Y., Chang, J. H., Koyama, T., Sato, M., ... Tsutsui, K. (2015). 7a-Hydroxypregnenolone, a key neuronal modulator of locomotion, stimulates upstream migration by means of the dopaminergic system in salmon. *Scientific Reports*, *5*(1), 12546. doi:10.1038/srep12546 PubMed

Harris, J., & Bird, D. J. (2000). Modulation of the fish immune system by hormones. *Veterinary Immunology and Immunopathology*, *77*(3-4), 163–176. doi:10.1016/S0165-2427(00)00235-X PubMed

Hirano, T. (1986). The spectrum of prolactin action in teleosts. In C. H. Ralpha (Ed.), *Comparative Endocrinology: Developments and Directions* (pp. 54–74). New York: Willey-Liss.

Ikuta, K. (1996). Effects of steroid hormones on migration of salmonid fishes. *Bulletin of National Research Institute of Aquaculture*, *2*(Supplement), 23–27.

Iwata, M., Tsuboi, H., Yamashita, T., Amemiya, A., Yamada, H., & Chiba, H. (2003). Function and trigger of thyroxine surge in migrating chum salmon Oncorhynchus keta fry. *Aquaculture (Amsterdam, Netherlands)*, *222*, 315–329. doi:10.1016/S0044-8486(03)00130-3

Kaneko, T., Watanabe, S., & Lee, K. M. (2008). Functional morphology of mitochondrion-rich cells in euroyhaline and stenohaline teleost. *Aqua-BioScience Monographs*, *1*(1), 1–62. doi:10.5047/absm.2008.00101.0001

Kindler, P. W., Philipp, D. P., Gross, M. T., & Bahr, J. M. (1989). Serum 11-ketotestosterone and testosterone concentrations associated with reproduction in male bluegill (Lepomis macrochirus: Centrarchidae). *General and Comparative Endocrinology*, *75*(3), 446–453. doi:10.1016/0016-6480(89)90180-9 PubMed

Kitahashi, T., Alok, D., Ando, H., Kaeriyama, M., Zohar, Y., Ueda, H., & Urano, A. (1998b). GnRH analog stimulate gonadotropin II gene expression in maturing sockeye salmon. *Zoological Science*, *15*(5), 761–765. doi:10.2108/zsj.15.761

Kitahashi, T., Ando, H., Ban, M., Ueda, H., & Urano, A. (1998a). Changes in the levels of gonadotropin subunit mRNAs in the pituitary of pre-spawning chum salmon. *Zoological Science*, *15*(5), 753–760. doi:10.2108/zsj.15.753

Kitahashi, T., Sato, A., Alok, D., Kaeriyama, M., Zohar, Y., Yamauchi, K., ... Ueda, H. (1998c). Gonadotropin-releasing hormone analog and sex steroids shorten homing duration of sockeye salmon in Lake Shikotsu. *Zoological Science*, *15*(5), 767–771. doi:10.2108/zsj.15.767

Kudo, H., Hyodo, S., Ueda, H., Hiroi, O., Aida, K., Urano, A., & Yamauchi, K. (1996). Cytophysiology of gonadotropin-releasing-hormone neurons in chum salmon (Oncorhynchus keta) forebrain before and after upstream migration. *Cell and Tissue Research*, *284*(2), 261–267. doi:10.1007/s004410050586 PubMed

Larsen, D. A., Swanson, P., & Dickhoff, W. W. (2011). The pituitary-thyroid axis during the parr-smolt transformation of coho salmon, Oncorhynchus kisutch: Quantification of TSHb mRNA, TSH, and thyroid hormones. *General and Comparative Endocrinology*, *171*(3), 367–372. doi:10.1016/j.ygcen.2011.03.003 PubMed

Lema, S. C., & Nevitt, G. A. (2004). Evidence that thyroid hormone induces olfactory cellular proliferation in salmon during a sensitive period for imprinting. *The Journal of Experimental Biology*, *207*(19), 3317–3327. doi:10.1242/jeb.01143 PubMed

McCormick, S. D. (2009). Evolution of the hormonal control of animal performance: Insights from the seaward migration of salmon. *Integrative and Comparative Biology*, *49*(4), 408–422. doi:10.1093/icb/icp044 PubMed

Munakata, A. (2012). Migratory behaviors in masu salmon (Oncorhynchus masou) and the influence of endocrinological factors. *Aqua-BioScience Monographs*, *5*(2), 29–65. doi:10.5047/absm.2012.00502.0029

Munakata, A., Amano, M., Ikuta, K., Kitamura, S., & Aida, K. (2001a). The involvement of sex steroid hormones in downstream and upstream migratory behavior of masu salmon. *Comparative Biochemistry and Physiology Part B*, *129*(2-3), 661–669. doi:10.1016/S1096-4959(01)00365-7 PubMed

Munakata, A., Amano, M., Ikuta, K., Kitamura, S., & Aida, K. (2001b). The effects of testosterone on upstream migratory behavior in masu salmon, Oncorhynchus masou. *General and Comparative Endocrinology*, *122*(3), 329–340. doi:10.1006/gcen.2001.7646 PubMed

Nagahama, Y. (1997). 17 Alpha; 20beta-dihydroxy-4-pregnen-3-one, a maturation-inducing hormone in fish oocytes: Mechanisms of synthesis and action. *Steroids*, *62*(1), 190–196. doi:10.1016/S0039-128X(96)00180-8 PubMed

Nagahama, Y., & Adachi, S. (1985). Identification of a maturation-inducing steroid in a teleost, the amago salmon (Oncorhynchus rhodurus). *Developmental Biology*, *109*, 428–435. doi:10.1016/0012-1606(85)90469-5

Ojima, D., & Iwata, M. (2007). The relationship between thyroxine surge and onset of downstream migration in chum salmon Oncorhynchus keta fry. *Aquaculture (Amsterdam, Netherlands)*, *273*(2-3), 185–193. doi:10.1016/j.aquaculture.2007.10.024

Ojima, D., & Iwata, M. (2009). Central administration of growth hormone-releasing hormone triggers downstream movement and schooling behavior of chum salmon (Oncorhynchus keta) fry in an artificial stream. Comparative Biochemistry and Physiology Part A, 152(3), 293–298. doi:10.1016/j.cbpa.2008.06.037

Ojima, D., & Iwata, M. (2010). Central administration of growth hormone-releasing hormone and corticotropin-releasing hormone stimulate downstream movement and thyroxine secretion in fall-smolting coho salmon (Oncorhynchus kisutch). *General and Comparative Endocrinology, 168*(1), 82–87. doi:10.1016/j.ygcen.2010.04.007 PubMed

Onuma, T., Higashi, Y., Ando, H., Ban, M., Ueda, H., & Urano, A. (2003). Year-to-year differences in plasma levels of steroid hormones in pre-spawning chum salmon. *General and Comparative Endocrinology, 133*(2), 199–215. doi:10.1016/S0016-6480(03)00171-0 PubMed

Pankhurst, N. W., & Barnett, C. W. (1993). Relationship of population density, territorial interaction and plasma levels of gonadal steroids in spawning male demoiselles Chromis dispulis (Pisces: Pomacentridae). *General and Comparative Endocrinology, 90*(2), 168–176. doi:10.1006/gcen.1993.1071 PubMed

Power, D. M. (2005). Developmental ontogeny of prolactin and its receptor in fish. *General and Comparative Endocrinology, 142*(1-2), 25–33. doi:10.1016/j.ygcen.2004.10.003 PubMed

Saito, Y., Mekuchi, M., Kobayashi, N., Kimura, M., Aoki, Y., Masuda, T., ... Yanagisawa, T. (2011). Molecular cloning, molecular evolution and gene expression of cDNAs encoding thyrotropin-releasing hormone receptor subtypes in a teleost, the sockeye salmon (Oncorhynchus nerka). *General and Comparative Endocrinology, 174*(2), 80–88. doi:10.1016/j.ygcen.2011.07.011 PubMed

Sanyal, T., & Sen, K. (2018). Diversified role of prolactin in fish: A review. *Global Journal of Engineering Science and Researches, 5*(11), 212–220. doi:10.5281/zenodo.1495605

Sato, A., Ueda, H., Fukaya, M., Kaeriyama, M., Zohar, Y., Urano, A., & Yamauchi, K. (1997). Sexual differences in homing profiles and shortening of homing duration by gonadotropin-releasing hormone analog implantation in lacustrine sockeye salmon (Oncorhynchus nerka) in Lake Shikotsu. *Zoological Science, 14*(6), 1009–1014. doi:10.2108/zsj.14.1009

Ueda, H. (1999). Artificial control of salmon homing migration and its application to salmon propagation. *Bulletin of Tohoku National Fisheries Research Institute, 62*, 133–139.

Ueda, H. (2011). Physiological mechanisms of homing migration in Pacific salmon from behavioral to molecular biological approaches. *General and Comparative Endocrinology, 170*(2), 222–232. doi:10.1016/j.ygcen.2010.02.003 PubMed

Ueda, H., Hiroi, O., Hara, A., Yamauchi, K., & Nagahama, Y. (1984). Changes in serum concentrations of steroid hormone, thyroxine, and vitellogenin during spawning migration of chum salmon, Oncorhynchus keta. *General and Comparative Endocrinology, 53*, 203–211. doi:10.1016/0016-6480(84)90243-0

Ueda, H., Nakamura, S., Nakamura, T., Inada, K., Okubo, T., Furukawa, N., ... Watanabe, M. (2016). Involvement of hormones in olfactory imprinting and homing in chum salmon. *Scientific Reports, 6*(1), 21102. doi:10.1038/srep21102 PubMed

Villars, T. A. (1983). Hormones and aggressive behavior in teleost fishes. In B. B. Svare (Ed.), Hormones and Aggressive Behavior (pp. 407–433). New York: Plenum Press; doi:10.1007/978-1-4613-3521-4_16.

Yamada, H., Amano, M., Okuzawa, K., Chiba, H., & Iwata, M. (2002). Maturational changes in brain contents of salmon GnRH in rainbow trout as measured by a newly developed time-resolved fluoroimmunoassay. *General and Comparative Endocrinology, 126*(2), 136–143. doi:10.1006/gcen.2002.7791 PubMed

Yamaguchi, S., Aoki, N., Kitajima, T., Iikubo, E., Katagiri, S., Matsushima, T., & Homma, K. (2012). Thyroid hormone determines the start of the sensitive period of imprinting and primes later learning. *Nature Communications, 3*(1), 1081. doi:10.1038/ncomms2088 PubMed

Yamauchi, K., Ban, M., Kasahara, N., Izumi, T., Kojima, H., & Harako, T. (1985). Physiological and behavioral changes occurring during smoltification in the masu salmon, Oncorhynchus masou. *Aquaculture (Amsterdam, Netherlands), 45*(1-4), 227–236. doi:10.1016/0044-8486(85)90272-8

Zohar, Y. (1996). New approaches for the manipulation of ovulation and spawning in farmed fish. *Bulletin of National Research Institute of Aquaculture, 2*(Supplement), 43–48.

Zohar, Y., & Mylonas, C. C. (2001). Endocrine manipulations of spawning in cultured fish: From hormones to genes. *Aquaculture (Amsterdam, Netherlands)*, *197*(1-4), 99–136. doi:10.1016/S0044-8486(01)00584-1

Chapter 3
Olfaction in Salmon

ABSTRACT

Olfaction is the most primitive and important chemosensory system for detecting a wide variety of environmental changes and is involved in various functions required for survival. Salmon have an efficient olfactory system; their olfactory abilities are inferior to those of dogs but superior to those of humans. Salmon olfactory systems can discriminate seasonally and yearly stable compositions of dissolved free amino acids in their natal streams produced by biofilms in the riverbed, and salmon can imprint on single amino acids before and during smoltification and discriminate these amino acids in the spawning season several years later. This chapter describes the structure and functions of the olfactory systems in salmon, previous olfactory hypotheses for salmon homing, properties of natal stream odorants, and biochemical and molecular biological studies on salmon olfactory functions.

INTRODUCTION

Olfaction, which is the most primitive sense of organisms, is the physiological capacity to detect various types of environmental information required for survival by chemoreception. More than 500 million years ago, organisms developed chemosensory systems to detect a wide variety of environmental changes (Hara, 1992). There are five essential sensor systems in vertebrates: audition (hearing), vision (seeing), somatosensation (touching), gustation (tasting), and olfaction (smelling). The numbers of receptor genes for each sensor are range from 1 for audition, to 3~4 for vision, 9 for somatosensation,

DOI: 10.4018/978-1-7998-2054-3.ch003

20~30 for gustation, and 150~1,500 for olfaction, indicate that olfaction can detect the most forms of environmental information. Fish living in water have both olfactory and gustatory systems to detect water-soluble cues. While gustation mainly involves the tasting of foods, olfaction is involved in a variety of functions: food recognition, crisis avoidance (alarm action), individual identification, kin recognition, intraspecies (conspecific) recognition, dominance hierarchies, symbiotic behavior, territorial behavior, schooling behavior, sexual behavior, and migratory behavior. As described in the previous chapter, hormones are secreted to adapt to environmental changes and to maintain homeostasis in cooperation with the ANS. Olfactory stimuli cause changes in hormone conditions. Conversely, hormonal changes also influence olfactory functions.

Structure and Function of the Olfactory System in Salmon

Since salmon respirate using gills, their nostrils have no function in respiration. Salmon possess an efficient water current system when swimming in which water flows from the anterior nostril to the posterior nostril (Figure 1A). In the nasal cavity, there is a so-called olfactory rosette, consisting of leaf-like structure of lamella resembling a miniature rose (Figure 1B). The numbers of the olfactory lamella is 3-4 in alevins, 11 in juveniles and 18 in adults of chum salmon (Yamamoto & Ueda, 1977; Kalinina et al., 2005). The olfactory receptor neuron (ORN) in the olfactory lamella is a bipolar neuron with a dendrite that detects water-soluble odorants and has an axon (i.e., the olfactory nerve: ON) that forms a synaptic contact with the dendrite of the mitral cell in the glomerulus of the OB (Figure 1C).

Kudo et al. (2009) examined the numbers of ORNs during the life cycle of chum salmon by means of morphometric analyses and revealed 0.18 and 14.2 million cells in fry and mature chum salmon, respectively. Dogs are well known for their well-developed olfactory ability to detect volatile odorants at low concentrations and are reported to have 200 million ORNs (Menco, 1989). The number of the ORNs in humans were reported to be 3 million (Moran et al., 1982). It is difficult to directly compare olfactory ability to detect water-soluble and volatile odorants between salmon and mammals on the basis of the numbers of ORNs; however, it is not difficult to infer that the olfactory abilities of salmon are inferior to those of dogs but superior to those of humans.

Figure 1. Olfactory systems in salmon. Lateral view of the salmon head showing the anterior and posterior nostrils (A), olfactory rosette showing the anterior and posterior nostrils, water current, and olfactory lamella (B), and the olfactory epithelium, olfactory nerve, olfactory receptor neuron in the olfactory lamella, and mitral cell and glomerulus in the olfactory bulb (C).

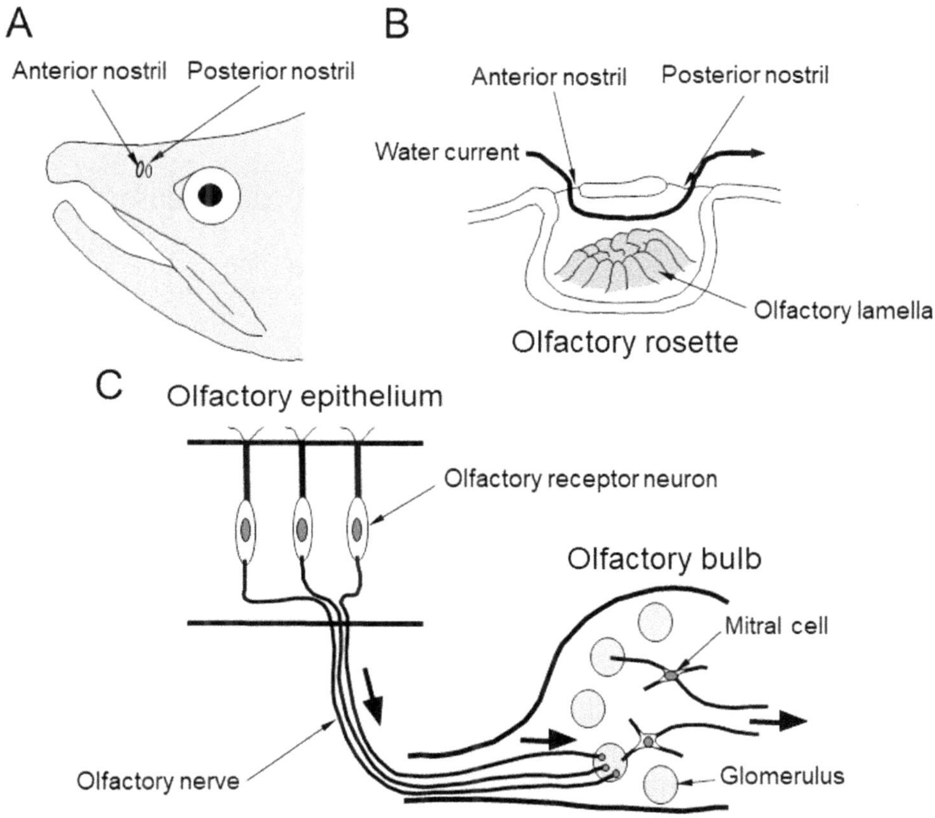

The mechanisms of olfaction are common in all vertebrates. Odorant substances binding to the olfactory receptor (OR) on the dendrite of the ORN activate G-protein-coupled adenylate cyclase, synthesize cAMP (G_{olf}. cAMP), open the cyclic nucleotide-gated ion channel, induce an influx of cations (Ca^{2+} and Na^{+}), and finally generate electric signals (Figure 2). There are three electrophysiological techniques for detecting electric signals produced in the ORN and transduced to the OB via the ON: electro-olfactograms (EOGs), ON recordings (ONRs), and electroencephalogram (EEGs). The

Figure 2. Mechanisms of olfaction in the olfactory receptor neuron (ORN). Odorant substances binding to the olfactory receptor on the dendrite of the ORN activate G-protein coupled adenylate cyclase, synthesize cAMP, open the cyclic nucleotide-gated ion channel, induce an influx of cations, and finally generate electric signals.

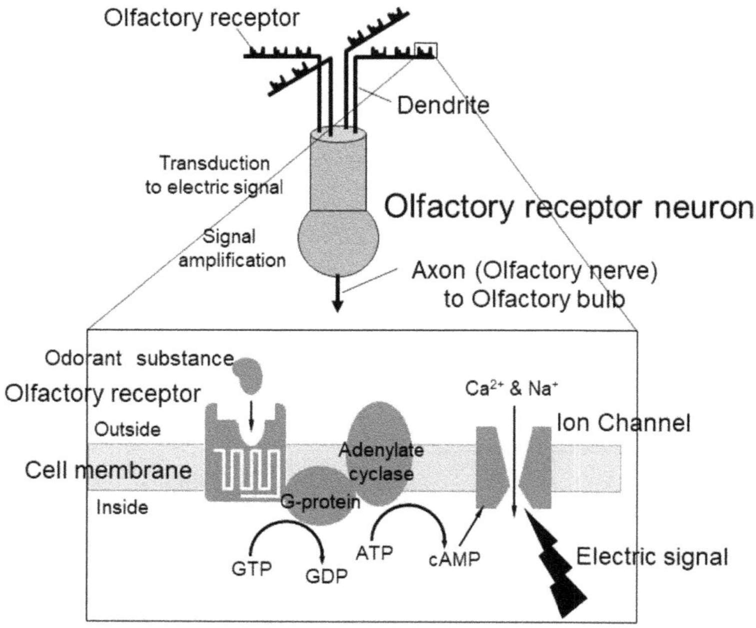

EOG records the summation of receptor potentials from the surface of the ORNs. The ONR represents the integration of action potentials from the bundle of the ON. The EEG records rhythmic oscillatory responses from the surface of the OB.

The number of OR genes varies greatly among vertebrate species, ranging from ~1,200 in rats to ~400 in humans, ~150 in zebrafish, and 15 in puffer fish (Niimura, 2009). In fish, there are three types of ORNs with different morphological shapes: ciliated ORNs, microvillous ORNs, and crypt cells, and the axon of each type converges to a specific region of the OB (Hamdani & Døving, 2007). Four major classes of chemicals were identified as specific olfactory stimuli, and their stimulatory effectiveness was characterized in fish: amino acids (a wide range), steroid hormones (DHP), bile salts, and prostaglandins ($PGF_{2\alpha}$) at thresholds of 10^{-9}~10^{-7} M, 10^{-12} M, 10^{-13}~10^{-12} M, and 10^{-13}~10^{-10} M, respectively (Hara, 1994).

Olfactory chemoreception is accomplished via the binding of an odorant substance to the OR, which is reportedly encoded by approximately 150 genes in the olfactory epithelium, with subsequent propagation of the information to the central nervous system in teleost fish (Alioto & Ngai, 2005). There are two types of *or* genes: the main olfactory receptors (MORs), which are expressed in ciliated olfactory receptor cells, and vomeronasal olfactory receptors (VORs), which are expressed in microvillous olfactory receptor cells (Hino et al., 2009). *mor* genes have also been identified in several salmonids (Wickens et al., 2001; Dukes et al., 2004, 2006; Morinishi et al. 2007). *or* gene expression was investigated in different life stages of Atlantic salmon, demonstrating that seven V2R-like (*olfc*) genes were expressed at higher levels in juveniles (parr and smolts) than in adults (Johnstone et al., 2011, 2012). Although many MORs and VORs have been identified from several vertebrate species owing to the progress of whole-genome analysis, many ligands remain uncharacterized. Further intensive molecular biological studies must be performed to clarify olfactory chemoreception during imprinting and homing migration in salmon.

Olfactory Hypotheses for Salmon Homing

To date, three different olfactory hypotheses have been proposed to explain how salmon home to their natal stream. The first hypothesis is the olfactory imprinting hypothesis, which was proposed by Wisby & Hasler (1954) using coho salmon; this hypothesis states that the specific odorant of the natal stream is imprinted in the olfactory system before juvenile salmon migrate to the sea and that adult salmon retrieve this odorant information as a cue for homing when they migrate through the home-stream network back to their home tributary (Hasler & Scholz, 1983). Harden-Jones (1968) developed the sequential imprinting hypothesis, which states that juvenile salmonids are imprinted in a sequential way at points located between the spawning area and the river mouth. The second hypothesis is known as the pheromone hypothesis, which was developed by Nordeng (1971, 1977) using Arctic char (*Salvelinus alpines*) and Atlantic salmon; this hypothesis states that homing adults are guided by population-specific odors or pheromones released from juvenile salmon. The pheromone hypothesis is not applicable to Pacific salmon because juveniles of chum and pink salmon are not present during the period when adults return. Finally, the third hypothesis is the hierarchical navigation hypothesis, which was recently proposed by Bett

& Hinch (2015); this hypothesis states that migrating adult salmon rely on imprinted odors as a primary directional cue, conspecific odors as a secondary directional cue, and nonolfactory environmental indicators as a tertiary directional cue. Among three hypotheses, the imprinting hypothesis, which states that certain specific odorant factors in the natal stream are imprinted on the olfactory system of juvenile salmon during downstream migration and that migrating adult salmon rely on these factors in the natal stream to attract them during the process of upstream homing migration, is the most applicable hypothesis to Pacific salmon (Dittman & Quinn, 1996).

Properties of Natal Stream Odors Detected by Salmon

In early electrophysiological studies on olfaction functions in salmon, the EEG response was used to identify natal stream odors, demonstrating that the application of natal stream water to the olfactory epithelium of homing coho salmon induced a large EEG response (Hara et al., 1965). However, it was later determined that this large EEG response was induced not only by natal stream waters but also by other stream waters (Hara, 1970). The natal stream odors were inferred to be absorbed by activated carbon and ion-exchange resin, insoluble in petroleum ether, dialyzable, nonvolatile, and heat-stable according to the spectral analysis of the EEG response of anadromous chum salmon and lacustrine sockeye salmon (Ueda, 1985).

In North America, artificial chemical odorant substances (β-phenylethyl alcohol (PEA) and morpholine) were used in imprinting experiments of coho salmon during smoltification, and these fish were then lured into unfamiliar streams scented with these odors during their homing migration several years later (Cooper et al., 1976; Scholz et al., 1976). The ORNs of adult coho salmon that were imprinted with PEA during smoltification were more sensitive to PEA than were those of nonimprinted fish (Nevitt et al., 1994). It was reported that it took 10 days for coho salmon exposed to PEA during smoltification to develop an imprinted memory (Dittman et al., 1996); additionally, the stimulation of guanylyl cyclase activity by PEA was significantly greater in olfactory cilia isolated from PEA-imprinted fish than in PEA-naïve fish at the time during the homing migration, which occurred 2 years after exposure to PEA (Dittman et al., 1977). Recently, experimental evidence of olfactory imprinting to a mixture of odorants containing amino acids and PEA in sockeye salmon was reported to occur at the embryonic and smolt stages (Havey et al., 2017). However, there have been few attempts to identify the properties of natal stream odors in salmon populations of the eastern Pacific.

The specific odors of each natal stream of salmon should be unique and distinct, should not vary between the spawning ground and the stream mouth, and should be stable during the years between juvenile imprinting and adult homing. The compositions of dissolved free amino acids (DFAA) and bile acids were analyzed in waters from streams that flow into Lake Toya (Shoji et al., 2000). The ONR response in cross-adaptation experiments of lacustrine masu salmon to artificial stream water (ASW) induced by the composition of DFAA resulted in strong responses, but the application of bile acid mixtures induced only very small responses. In cross-adaptation experiments, the ONR pattern for each combination of ASW closely resembled that of the corresponding combination of natural stream water (Figure 3). These results indicated that the composition of DFAA in the natal stream water represents the natal stream odors recognized by masu salmon.

The ONR responses of lacustrine masu salmon and sockeye salmon were examined to determine their olfactory discrimination ability for various freshwaters; the results confirmed different responses were elicited in the two species, regardless of sex or gonadal maturity (Sato et al., 2000). In cross-adaptation experiments, the secondary response to stream water was abolished by lake water; however, the secondary response to lake water was

Figure 3. Typical integrated olfactory nerve recording (ONR) responses to natural stream waters and artificial stream waters reconstituted with amino acids and salts based on the compositions of natural stream waters after the olfactory epithelium was adapted to the lake water.

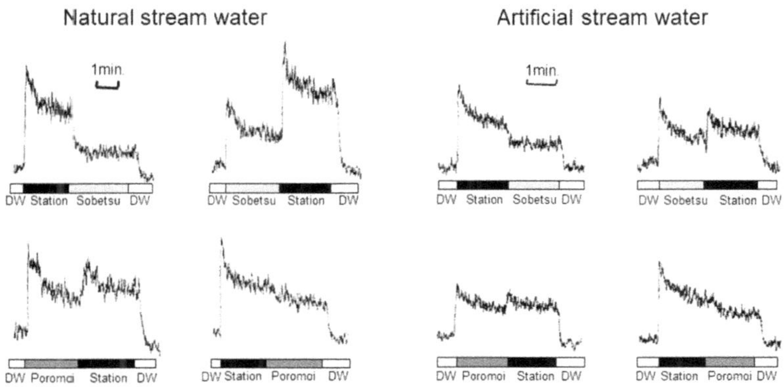

not abolished by stream water. This result is reasonable because both species migrate from lake to stream. The threshold (i.e., minimum concentration) that induced the ONR response to stream water after adaptation to lake water occurred at between 0.1 and 1.0% in lacustrine sockeye salmon and masu salmon, revealing that the olfactory discriminatory ability is effective only within a limited distance from the natal stream.

In watershed ecosystems, complicated biological processes, such as soil, vegetation, litter, pollen, dew, and various microbial activities, cause changes in the composition of DFAA (Thomas, 1997). Moreover, the roles of complex microbial communities known as biofilms, which consist of various microorganisms that are embedded in a matrix of extracellular polymeric substances, have been intensively studied (Costerton et al., 1994; Nosyk et al., 2008). The origin of DFAA in stream water was investigated by focusing on biofilms in a stream bed via incubation experiments in the laboratory (Ishizawa et al., 2010). Three stones were placed in the Toyohira River for 3 months to allow the formation of biofilms and were then incubated in the laboratory at stream water temperature for 24 hours. The composition

Figure 4. Average concentration (nM) and composition (mol %) of dissolved free amino acids in the incubation solution of the biofilm experiment at 0 hours (initial control) and 24 hours (N=3).

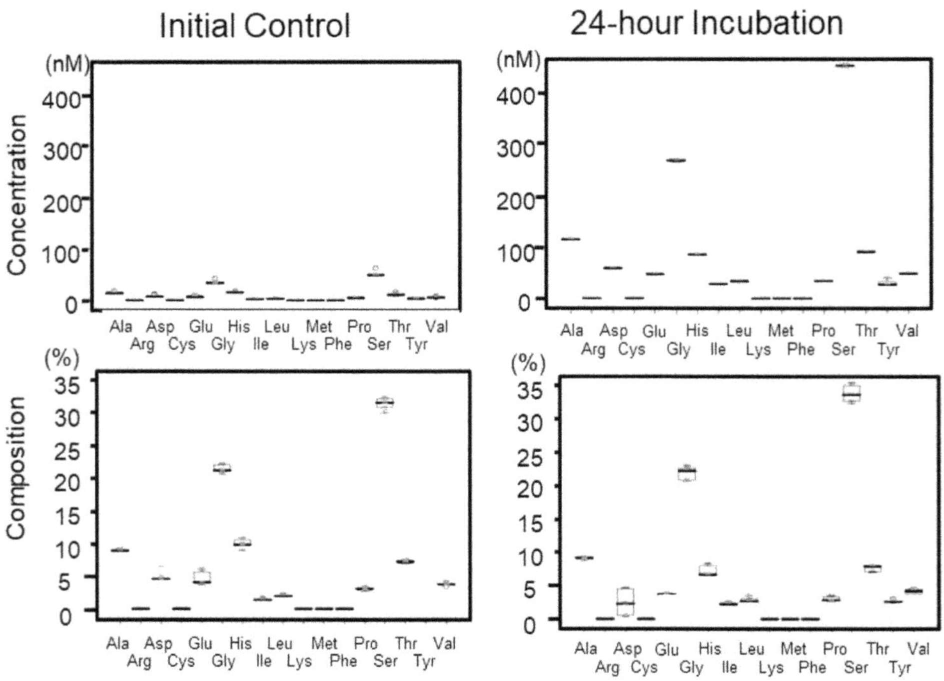

and concentration of DFAA in the incubation solution were measured after incubation. The total average concentration of DFAA increased greatly in the biofilm incubation solution (initial control: 162.6±7.6 nM; 24-hour incubation: 812.9±247.1 nM). However, the average composition of DFAA showed little change after 24 hours of incubation (Figure 4). These results revealed that biofilms are a primary source of DFAA in stream water.

Comparison of Olfactory Discrimination Abilities Among Four Pacific Salmon Species

The upstream selective movements of the four Pacific salmon species (i.e., pink salmon, chum salmon, sockeye salmon, and masu salmon) were

*Figure 5. Behavioral selectivity for the artificial natal stream waters of mature males of four Pacific salmon species (masu, sockeye, chum, and pink salmon) in the two-choice test tank. The numbers in boxes at the top indicate the number of fishes travelled upstream into either arm of the tank/the total number of fishes experimentally tested, showing the % of upstream movement. The numbers in parenthesis indicate the number of fishes. Data are presented as the means ± SEM. Significant differences are indicated at *p<0.05 level.*

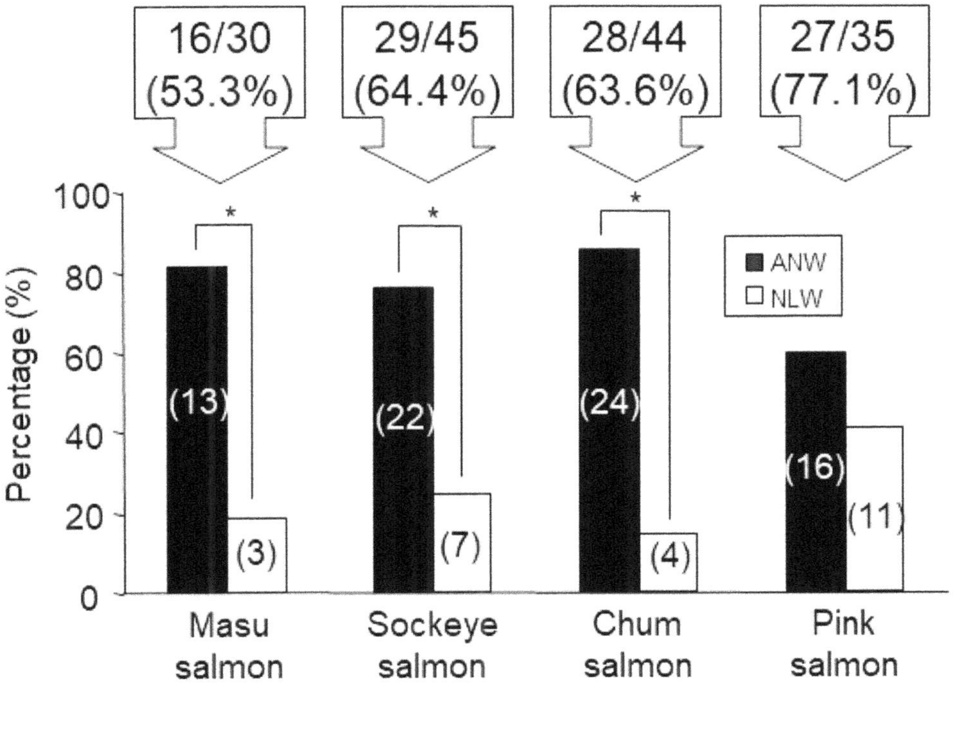

compared using artificial natal stream water (ANW) that was prepared to exhibit the same composition and concentration of DFAA as those found in their natural natal streams, and the experiment took place in a 2-choice test tank (i.e., a Y-maze) that consisted of two water inlet arms and one pool. Two different types of water (ANW and control lake water) were added to the water inlet of either the left or the right arm, and fish movement was monitored to determine the number of fishes that entered each arm. Among the four Pacific salmon species, pink salmon showed the highest percentage of upstream movement but showed the lowest level of selectivity for the arm with ANW (Figure 5). However, the other three Pacific salmon species (i.e., chum salmon, sockeye salmon, and masu salmon) showed significant selectivity for the arm with ANW (Yamamoto et al., 2008; Ueda, 2011).

If salmon always accurately home to the streams where they were born, there will be a potential for reduced genetic diversity along with a restricted distribution and small population size. Thus, it is quite interesting to investigate the relationship between salmon evolution and homing accuracy from an evolutionary perspective. Compared to pink salmon, masu salmon have a restricted distribution and small population size (Kaeriyama & Ueda, 1998) (Figure 3B in Chapter 1). These fish also return to their natal stream with the highest precision, whereas pink salmon are more likely to stray into a nonnatal stream. Pink salmon may have evolved the capacity to adapt to nonnatal stream odors, enabling this expansion in their distribution. Retropositional genome analyses have revealed that pink salmon are phylogenetically the most advanced salmon species, while masu salmon are a more primitive species (Murata et al., 1996) (Figure 3A in Chapter 1).

Olfactory Discrimination Abilities of Chum Salmon

The olfactory discrimination of stream odors in juvenile chum salmon was measured during downstream migration from the Chitose Hatchery in Hokkaido, Japan (Site **A**) to Ishikari Bay (Site **F**) (Figure 5 in Chapter 2) using EOG. Fish were collected at the second bridge of the Chitose (Site B: the first collection site after release from the hatchery) and the confluence point of the Chitose and Old Yubari Rives (Site **D**). Fish that encountered streams with water that differed from that of the hatchery showed significantly greater responses to the Ishikari River water than to the Chitose and Toyohira River waters (Figure 6A) (Ueda et al., 2016). This phenomenon supports the sequential imprinting hypothesis developed by Harden-Jones (1968). The EOG responses of adult male chum salmon collected at the Indian waterwheel in the Chitose River (Site **J**) were significantly higher for both the Ishikari River and the Chitose River (i.e., their natal stream waters) than for the Toyohira River, which is a nearby tributary that is not part of

Figure 6. Electro-olfactogram (EOG) responses to waters from three streams in juvenile chum salmon during downstream migration from Site A to Site F (A). EOG responses to waters from three streams in adult male chum salmon collected at Site J (B). Data are presented as the means ± SEM (A, N=3; B, N=7). Different letters represent significant differences at p<0.05 level.

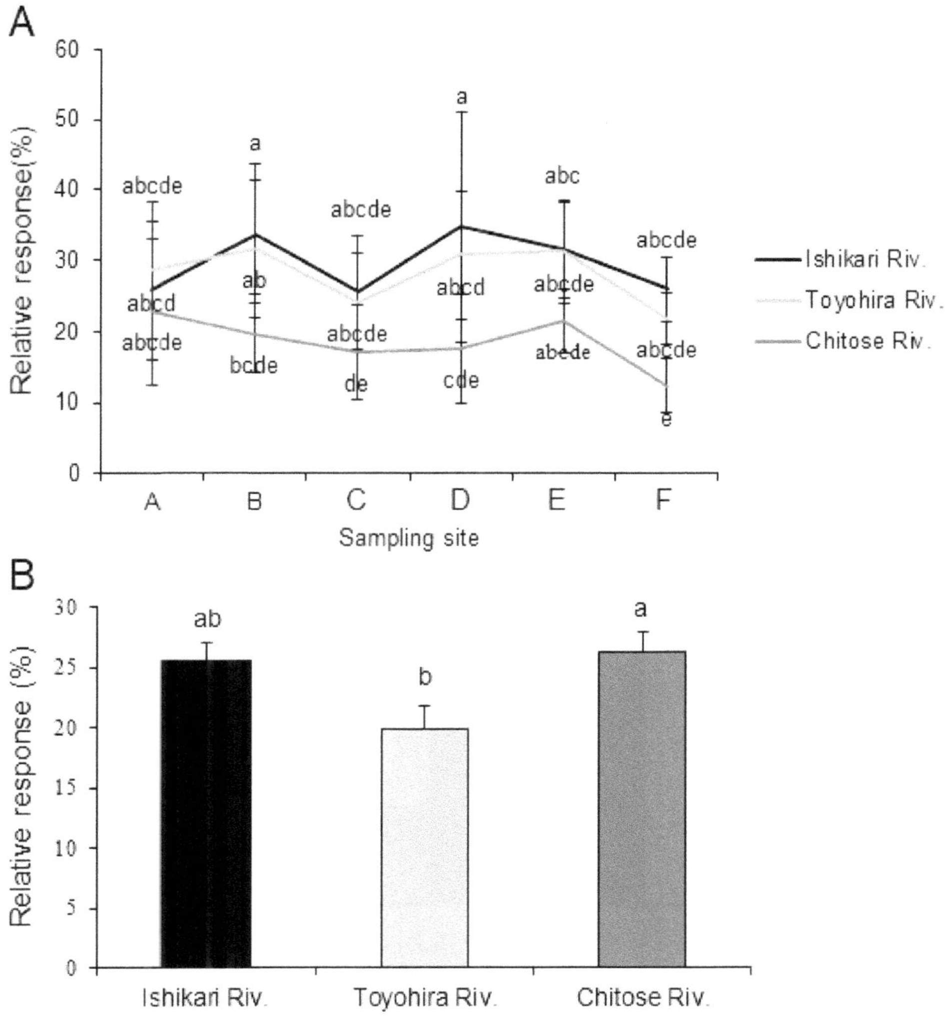

their migratory path (Figure 6B) (Ueda et al., 2016; Ueda, 2019). These data suggest that the olfactory memory retrieval abilities regarding natal stream odors increase during the upstream migration of adult chum salmon

Behavioral selection experiments of chum salmon captured in the Osaru River, Hokkaido, were also conducted in a Y-maze using various combinations of control water (NLW) and three artificial stream waters prepared with the same composition and concentration of DFAA found in natural stream waters: 1. artificial Osaru River water (AOR); 2. AOR without L-glutamic acid, the major amino acid in the Osaru River water (AOR-E); 3. artificial water matching another stream (ALS) that had much

*Figure 7. Behavioral selectivity for each artificial stream water (NLW: natural lake water; R: right arm; L: left arm; AOR: artificial Osaru River water; AOR-E: AOR without L-glutamic acid; ALS: artificial Lake Shikotsu water) in mature male chum salmon in the two-choice test tank. The numbers in boxes at the top indicate the number of fishes traveled upstream into either arm of the tank/the total number of fishes experimentally tested, showing the % of upstream movement. The numbers in parenthesis indicate the number of fishes. Data are presented as the means ± SEM. Significant differences are indicated at *p<0.05 level.*

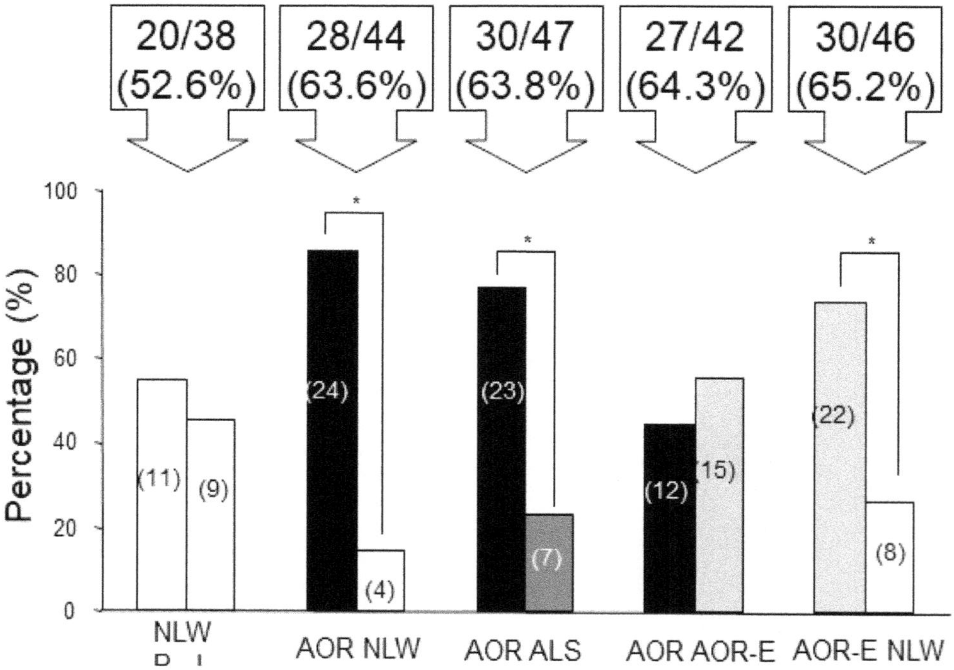

higher amino acid concentrations than the Osaru River (Yamamoto & Ueda, 2009). In the behavioral tests, the fish did not discriminate between AOR and AOR-E but displayed significant selection for AOR or AOR-E over NLW and AOR over ALS (Figure 7). The ONR response in cross-adaptation experiments indicated that mature male chum salmon have the olfactory ability to distinguish between AOR and AOR-E. These results suggest that migratory male chum salmon respond to DFAA mixtures in their natal stream water and appear not to be affected by single amino acids.

Figure 8. Seasonal and yearly changes in the composition of 15 dissolved free amino acids in the Teshio River. The stable compositions of seven amino acids (Ala, Asp. Glu, Gly, Leu, Ser, and Val) between May 2005 and September 2009 (A) and five amino acids (Ala, Asp, Glu, Ser, and Thr) between May 2006 and September 2010 (B) are indicated in black boxes.

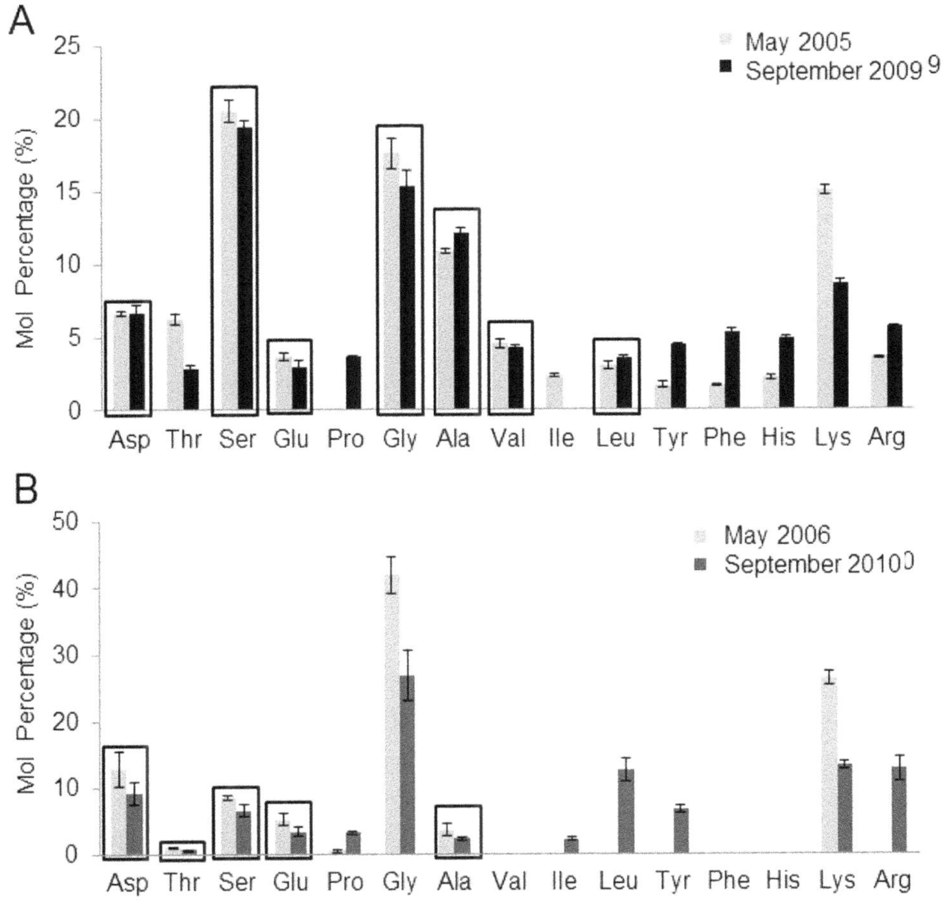

*Figure 9. Behavioral selectivity for each artificial stream water (NLW: natural lake water; R; right arm; L: left arm; jASW: artificial stream water during juvenile downstream migration in May 2005 and 2006; aASW, artificial stream water during adult upstream migration in September 2009 and 2010) in 4-year-old adult male chum salmon in the 2-choice test tank in 2009 and 2010. The numbers in boxes at the top indicate the number of fishes travelled upstream into either arm of the tank/the total number of fishes experimentally tested, showing the % of upstream movement. The numbers in parenthesis indicate the number of fishes. Data are presented as the means ± SEM. Significant differences are indicated at *p<0.05 level.*

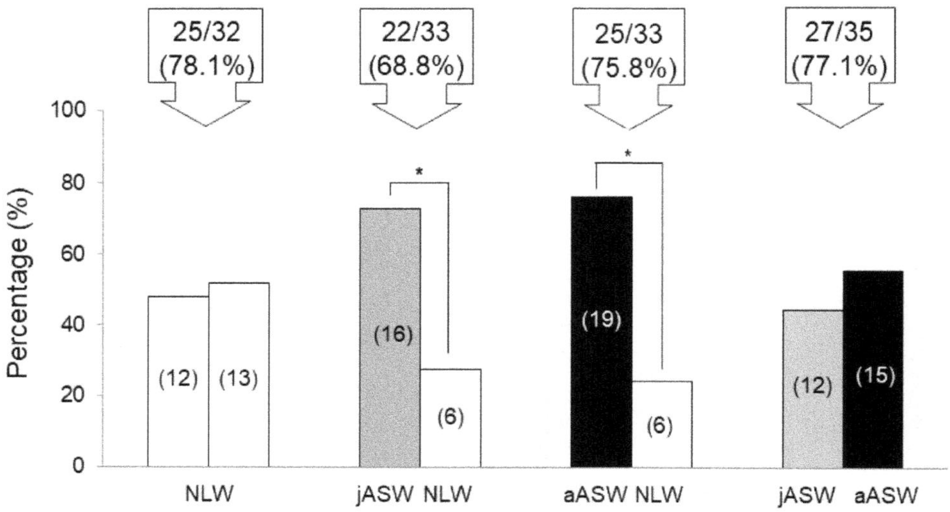

The concentration and composition of DFAA in stream water may change seasonally and annually. In the Teshio River in Hokkaido, Japan, juvenile chum salmon undergo their downstream migration in spring, and adults undergo their upstream migration in autumn; there is a 4-year span between the downstream and upstream migrations. The Teshio River water contained 15 DFAAs, and the concentrations of these DFAAs fluctuated greatly. However, the compositions (mole %) of 5~7 DFAAs were stable between the spring and autumn samples within a 4-year span (Figure 8).

The same concentrations of DFAA as those found in the Teshio River during the time of juvenile imprinting in spring (jASW) and adult homing in autumn (aASW) were prepared after a 4-year period. Behavioral experiments addressing upstream selective movement were conducted in a Y-maze (Yamamoto et al., 2013). Four-year-old mature male chum salmon

Figure 10. Effects of GnRH analog (GnRHa) implantation on the EOG response to three stream waters in adult male chum salmon collected from Ishikari Bay. Adult chum salmon including strays from another hatchery (A) and adult chum salmon originating from the Chitose Hatchery (B) caught at Site F. Data are presented as the means ± SEM (N=4-5). Different letters represent significant differences at p<0.05 level.

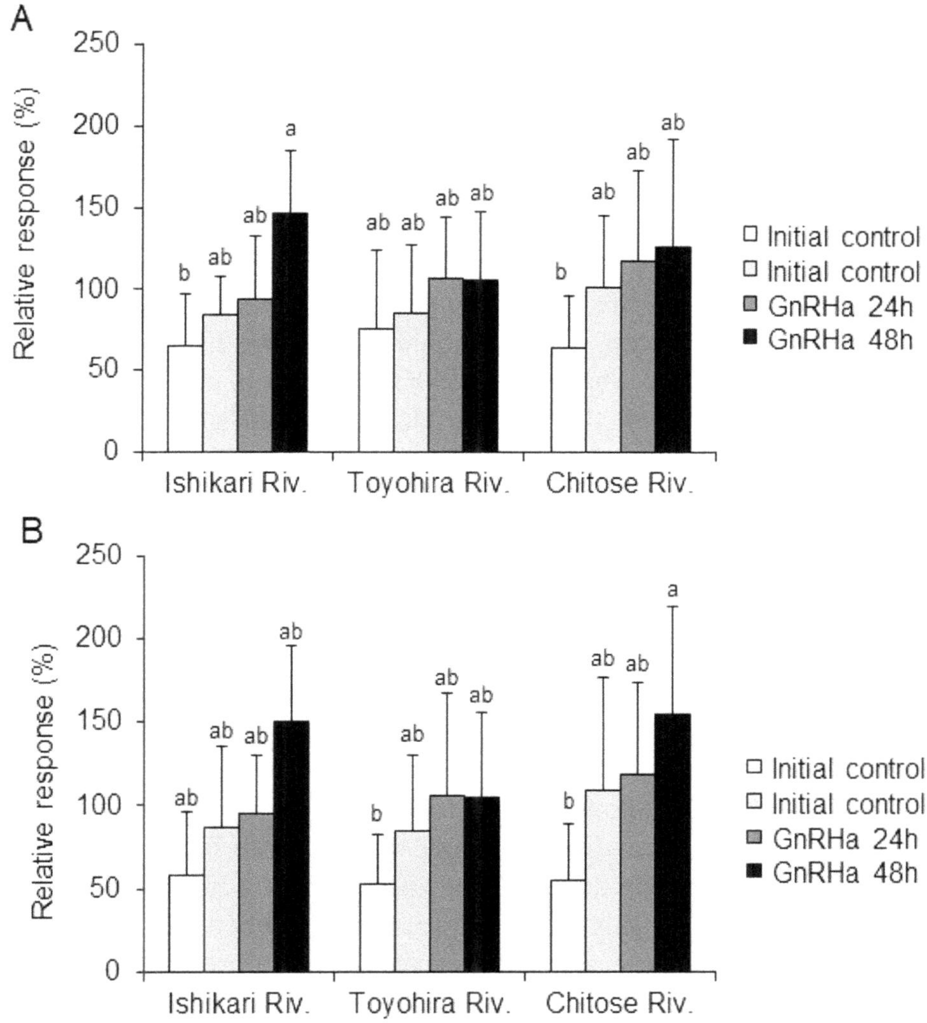

captured in the Teshio River showed a significant preference for either jASW or aASW over control water, but they did not display a preference between jASW and aASW (Figure 9). These results clearly confirm that the long-term stability of the composition of DFAA in the natal stream is crucial for olfactory imprinting and homing in chum salmon.

Furthermore, Chen et al. (2017) monitored the homing behaviors of chum salmon toward river-specific DFAA profiles using a passive integrated transponder (PIT) tag system in a Y-maze. Most chum salmon that homed to the Chitose River showed a preference for the Chitose River ASW and tended to stay in the corresponding arm of the maze for a longer period. These results confirmed the ability of chum salmon to identify and discriminate ASW that contained natural levels of DFAA.

The effects of implanting GnRHa on the olfactory discrimination ability for natal stream odors were studied using adult male chum salmon caught in Ishikari Bay (Site **F**, Figure 5 in Chapter 2) prior to entering the Ishikari River. GnRHa implantation increased the serum levels of DHP after 48 hours (Ueda et al. 2016; Ueda, 2019). GnRHa implantation for 48 hours significantly increased the relative EOG response to the Ishikari River water in adult male chum salmon, including straying fish from another hatchery (Figure 10A). The relative EOG response of adult male chum salmon originating from the Chitose Hatchery (confirmed by thermal otolith marks) showed significant discrimination of Chitose River water (Figure 10B), revealing that GnRHa stimulated the discriminating ability of natal stream odor.

Investigation of Artificial Imprinting using Amino Acids

Using electrophysiological and behavioral experiments, Yamamoto et al. (2010) revealed that 1-year-old lacustrine sockeye salmon can be imprinted at smoltification by a single amino acid, 1 μM L-proline (Pro) or L-glutamic acid (Glu). The EOG responses of test fish exposed to Pro in March (before smoltification) and April-June (during smoltification) for 2 weeks were significantly greater than those of nonexposed control fish, but not those of test fish exposed in July (after smoltification). When Pro and control water were added to the water inlets of the Y-maze during the spawning season 2 years after test water exposure, 80% of maturing and mature fish exposed before and during smoltification showed a preference for Pro, whereas those exposed after smoltification displayed no preference (Figure 11). The EOG response of fish exposed to Pro or Glu for 1 hour, 6 hours, 1 day, 7 days, or 14 days in May revealed that only the response after 14 days of exposure was significantly greater than that in the control. One-

*Figure 11. Behavioral selectivity of mature sockeye salmon among experimental fish (exposed to 1 μM L-proline from March to July in 2005) and unimprinted control fish in a two-choice tank containing either the experimental water (Pro) or natural lake water (NLW) in October 2007. The numbers in boxes at the top indicate the number of fishes that traveled upstream into either arm of the tank/the total number of fishes experimentally tested, showing the % of upstream movement. The numbers in parenthesis indicate the number of fishes. Data are presented as the means ± SEM. Significant differences are indicated at *p<0.05 level.*

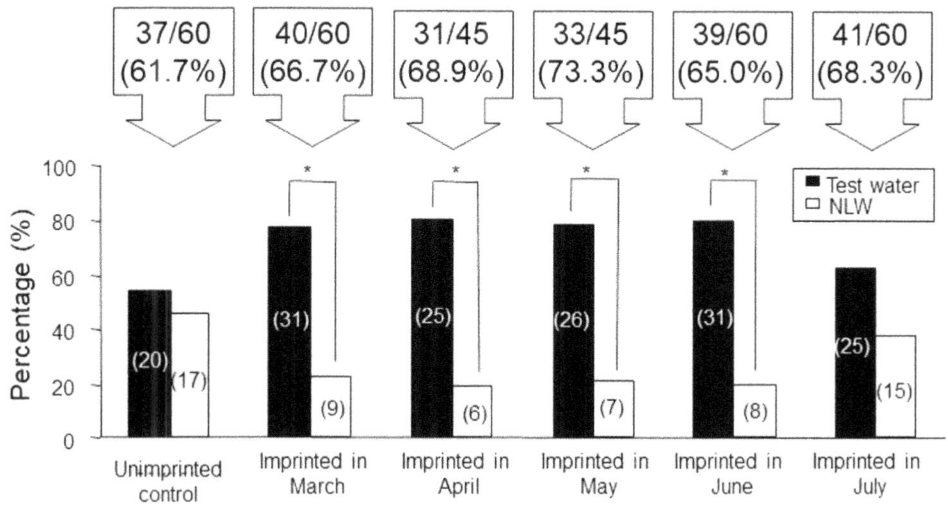

year-old sockeye salmon can imprint on a single amino acid before and during smoltification and this imprinting requires exposure for at least 14 days in the artificial rearing environment. In a natural stream environment, however, smolts should immediately imprint on different odors when they encounter a branch stream that flows into a main stream during downstream migration, as in the sequential imprinting hypothesis proposed by Harden Johns (1968). Further experiments should be designed to clarify how smolts are imprinted when they encounter an odor for a short time during downstream migration using an artificial raceway tank.

Biochemical and Molecular Biological Studies on Salmon Olfactory Functions.

An olfactory system-specific protein of 24 kDaltons (N24) was identified in lacustrine sockeye salmon by electrophoretic comparison of proteins restricted to the olfactory system with those found in other parts of the brain (Shimizu et al., 1993). In various species of teleosts, N24 immunoreactivity was found in the olfactory system of a catadromous species, Japanese eel (*Anguilla japonica*), but not in a nonmigratory species, carp (*Cyprinus carpio*) (Ueda et al. 1994). Interestingly, N24 immunoreactivity was also observed in the testicular germ cells, spermatids and spermatozoa, suggesting the involvement of N24 in sperm chemotaxis (Ueda et al. 1993). The cDNA encoding *n24* was isolated and sequenced, and this cDNA contained a coding region encoding 216 amino acid residues. Protein and nucleotide sequencing demonstrated the existence of remarkable homology between N24 and glutathione S-transferase class pi enzymes (Kudo et al., 1999). Northern blot analysis showed that *n24* mRNA with a length of 950 bases was expressed in the lacustrine sockeye salmon olfactory epithelium. The functional roles of N24 during salmon imprinting and homing migration should be clarified in future studies.

Copper can be an effective hatchery treatment for the control of certain parasitic and fungal infections in salmonids, yet it is also highly toxic to the olfactory system in fishes. Numerous reports have described the early inhibitory impacts of copper on fish olfaction, but very few have addressed the process of recovery. Sandahl et al. (2005) found that the chum salmon olfactory system could recover from short-term (4 h) exposure to copper (3-58 μg/L) within 1 day, as indicated by EOG recordings. Recovery followed initial EOG reductions of up to 88%. A novel cytophysiological approach was applied to monitor the changes in the OR cells using a fluorescent styryl dye, AMI-43. The influence of copper on the number of OR cells labeled with the dye was biphasic, initially increasing the number at the two lowest copper concentrations (3 and 8 μg/L) and decreasing the number at the two highest concentrations (24 and μg/L). Changes in the ability of OR neurons to internalize the dye were monitored for up to 10 days in fish recovering from copper exposure.

Hino et al. (2007) investigated the salmon olfactory imprinting-related gene (SOIG: *soig*) in the olfactory system of lacustrine sockeye salmon using the subtractive hybridization technique of cDNA-representational difference analysis using fish at smoltification as the test fish and fish in the feeding migration period as drivers. *soig* mRNA was shown to be expressed in olfactory receptor cells and basal cells of the olfactory epithelium by *in situ* hybridization. The expression levels of *soig* mRNA in the olfactory epithelium were analyzed during the smoltification and homing of sockeye salmon and chum salmon, respectively. During smoltification, *soig* mRNA levels increased before and during smoltification, and they decreased after smoltification in sockeye salmon. These changes coincided with serum thyroxine changes during smoltification (Yamamoto et al., 2010). During homing migration, *soig* mRNA levels in the olfactory epithelium of chum salmon were elevated at Site **F** and Site **J** (Figure 5 in Chapter 2). It is considered possible that SOIG might be related to olfaction or cell proliferation during both smoltification and the final stage of homing.

Palstra et al. (2015) aimed to provide insights into the hypothesized functional link between the olfactory sensing of the spawning ground and final sexual maturation. The presence and expression levels of olfactory genes were assessed by RNA sequencing (RNAseq) of the olfactory rosettes in homing chum salmon from the coastal sea to 75 km upstream of the rivers at the prespawning ground. The progression of sexual maturation along the BPG axis was assessed through the determination of plasma steroid levels by TR-FIA, pituitary *gth* subunit expression and *sgnrh* expression in the brain via quantitative real-time PCR. RNAseq revealed the expression of 75 known and 27 unknown salmonid olfactory genes, among which 13 genes were differentially expressed between fish from the prespawning area and the coastal area, suggesting an important role of these genes in homing. A clear progression towards final maturation was characterized by higher plasma DHP levels, greater numbers of receptor genes being expressed, increased pituitary *lhβ* expression and *sgnrh* expression in the post brain, and lower plasma T and E2 levels. Olfactomedins and ependymin are candidates among the differentially expressed genes that may connect olfactory reception to the expression of sgnrh to regulate final maturation.

CONCLUSION

Olfaction is the most primitive and important chemosensory system in organisms for detecting a wide variety of environmental changes and is characterized by the greatest number of receptor genes among the five essential sensor systems in vertebrates. Mechanisms of olfaction are common in all vertebrates, where odorant substances binding to the OR of the ORN generate electric signals transduced to the OB via the ON. Salmon possess efficient olfactory abilities to detect water-soluble odorants involved in various important functions: food recognition, crisis avoidance (alarm action), individual identification, kin recognition, intraspecies (conspecific) recognition, dominance hierarchy, symbiotic behavior, territorial behavior, schooling behavior, sexual behavior, and migratory behavior. Three olfactory hypotheses have been proposed to explain the homing of salmon to their natal stream, among which the olfactory imprinting hypothesis, which states that certain specific odorant factors in the natal stream are imprinted on the olfactory system of juvenile salmon during downstream migration and that migrating adult salmon rely on these factors in the natal stream to attract them during the process of upstream homing migration, is the most applicable hypothesis for Pacific salmon. The specific odors of each natal stream of salmon should be unique and distinct, should not vary between the spawning ground and the stream mouth, and should be stable during the years between juvenile imprinting and adult homing. The long-term stability of the composition of DFAA produced mainly by biofilms in riverbeds is crucial for olfactory imprinting and homing in chum salmon. Among four examined Pacific salmon species (pink, chum, sockeye, and masu salmon), pink salmon showed the highest percentage of upstream movement but the lowest level of selectivity for ANW prepared with the same composition and concentration of DFAA found in their natal stream, suggesting that pink salmon may have evolved the capacity to adapt to nonnatal stream odors to expand their distribution in the Pacific Ocean. It is necessary to investigate the relationship between salmon evolution and homing accuracy from an evolutionary perspective. The GnRHa implantation experiment in chum salmon prior to entering the natal stream revealed that GnRHa can stimulate the discrimination of the natal stream odor. One-year-old sockeye salmon could imprint on a single amino acid before and during smoltification in the artificial imprinting experiment; however, it should be investigated how BPT hormones might be involved in olfactory imprinting in Pacific salmon. Furthermore, there are important unresolved issues regarding the olfactory imprinting and discrimination of natal stream odorants between accurately homing salmon and straying salmon.

REFERENCES

Alioto, T. S., & Ngai, J. (2005). The odorant receptor repertoire of teleost fish. *BMC Genomics*, *6*(1), 173. doi:10.1186/1471-2164-6-173 PMID:16332259

Bett, N. N., & Hinch, S. G. (2015). Olfactory navigation during spawning migrations: A review and introduction of the Hierarchical Navigation Hypothesis. *Biological Reviews of the Cambridge Philosophical Society*, *91*(3), 728–759. doi:10.1111/brv.12191 PMID:25923999

Chen, E. Y., Leonard, J. B. K., & Ueda, H. (2017). The behavioural homing response of adult chum salmon *Oncorhynchus keta* to amino acid profiles. *Journal of Fish Biology*, *90*(4), 1257–1264. doi:10.1111/jfb.13225 PMID:27873320

Cooper, J. C., Scholz, A. T., Horrall, R. M., Hasler, A. D., & Madison, D. M. (1976). Experimental confirmation of the olfactory hypothesis with artificially imprinted homing coho salmon (*Oncorhynchus kisutch*). *Journal of the Fisheries Research Board of Canada*, *33*(4), 703–710. doi:10.1139/f76-087

Costerton, J. W., Lewandowski, Z., DeBeer, D., Caldwell, D., Korber, D., & James, G. (1994). Biofilms, the customized microniche. *Journal of Bacteriology*, *176*(8), 2137–2142. doi:10.1128/jb.176.8.2137-2142.1994 PMID:8157581

Dittman, A. W., & Quinn, T. P. (1996). Homing in Pacific salmon: Mechanisms and ecological basis. *The Journal of Experimental Biology*, *199*, 83–91. Retrieved from jeb.biologists.org/content/199/1/83 PMID:9317381

Dittman, A. W., Quinn, T. P., & Nevitt, G. A. (1996). Timing of imprinting to natural and artificial odours by coho salmon (*Oncorhynchus kisutch*). *Canadian Journal of Fisheries and Aquatic Sciences*, *53*(2), 434–442. doi:10.1139/f95-185

Dittman, A. W., Quinn, T. P., Nevitt, G. A., Hacker, B., & Storm, D. R. (1997). Sensitization of olfactory guanylyl cyclase to a specific imprinted odorant in coho salmon. *Neuron*, *19*(2), 381–389. doi:10.1016/S0896-6273(00)80947-2 PMID:9292727

Dukes, J. P., Deaville, R., Bruford, M. W., Youngson, A. F., & Jordan, W. C. (2004). Odorant receptor gene expression changes during the parr-smolt transformation in Atlantic salmon. *Molecular Ecology*, *13*(9), 2851–2857. doi:10.1111/j.1365-294X.2004.02252.x PMID:15315695

Dukes, J. P., Deaville, R., Gottelli, D., Neigel, J. E., Bruford, M. W., & Jordan, W. C. (2006). Isolation and characterisation of main olfactory and vomeronasal receptor gene families from the Atlantic salmon (*Salmo salar*). *Gene*, *371*(2), 257–267. doi:10.1016/j.gene.2005.12.003 PMID:16481129

Hamdani, E. H., & Døving, K. B. (2007). The functional organization of the fish olfactory system. *Progress in Neurobiology*, *82*(2), 80–86. doi:10.1016/j.pneurobio.2007.02.007 PMID:17433527

Hara, T. J. (1970). An electrophysiological basis for olfactory discrimination in homing salmon: A review. *Journal of the Fisheries Research Board of Canada*, *27*(3), 565–586. doi:10.1139/f70-059

Hara, T. J. (1992). Overview and Introduction: Mechanisms of Olfaction. In T.J. Hara (Ed.), Fish Chemoreception. London: Chapman and Hall.

Hara, T. J. (1994). The diversity of chemical stimulation in fish: Olfaction and gustation. *Reviews in Fish Biology and Fisheries*, *4*(1), 1–35. doi:10.1007/BF00043259

Hara, T. J., Ueda, K., & Gorbman, A. (1965). Electroencephalographic studies of homing salmon. *Science*, *149*(3686), 884–885. doi:10.1126cience.149.3686.884 PMID:14332854

Harden-Jones, F. R. (1968). *Fish Migration*. London: Edward Arnold.

Hasler, A. D., & Scholz, A. T. (1983). *Olfactory imprinting and homing in salmon*. Berlin: Springer-Verlag. doi:10.1007/978-3-642-82070-0

Havey, M. A., Dittman, A. H., Quinn, T. P., Lema, S. C., & May, D. (2017). Experimental evidence for olfactory imprinting by sockeye salmon at embryonic and smolt stages. *Transactions of the American Fisheries Society*, *146*(1), 74–83. doi:10.1080/00028487.2016.1238409

Hino, H., Iwai, T., Yamashita, M., & Ueda, H. (2007). Identification of an olfactory imprinting-related gene in the lacustrine sockeye salmon, *Oncorhynchus nerka. Aquaculture (Amsterdam, Netherlands)*, *273*(2-3), 200–208. doi:10.1016/j.aquaculture.2007.10.003

Hino, H., Miles, N. G., Bandoh, H., & Ueda, H. (2009). Molecular biological research on olfactory chemoreception in fishes. *Journal of Fish Biology*, *75*(5), 945–959. doi:10.1111/j.1095-8649.2009.02341.x PMID:20738593

Ishizawa, S., Yamamoto, Y., Denboh, T., & Ueda, H. (2010). Release of dissolved free amino acids from biofilms in stream water. *Fisheries Science*, *76*(4), 669–676. doi:10.100712562-010-0246-2

Johnstone, K. A., Lubienieck, K. P., Koop, B. F., & Davidson, W. S. (2011). Expression of olfactory receptors in different life stages and life histories of wild Atlantic salmon (*Salmo salar*). *Molecular Ecology*, *20*(19), 4059–4069. doi:10.1111/j.1365-294X.2011.05251.x PMID:21883590

Johnstone, K. A., Lubienieck, K. P., Koop, B. F., & Davidson, W. S. (2012). Identification of olfactory receptor genes in Atlantic salmon *Salmo salar. Journal of Fish Biology*, *81*(2), 559–575. doi:10.1111/j.1095-8649.2012.03368.x PMID:22803724

Kaeriyama, K., & Ueda, H. (1998). Life history strategy and migration pattern of juvenile sockeye (*Oncorhynchus nerka*) and chum salmon (*O. keta*) in Japan: A review. *North Pacific Anadromous Fish Commission Bulletin*, *1*, 163–171.

Kalinina, C. G., Matrosova, I. V., Doroshenko, M. A., & Evdokimov, W. (2005). Morphohistochemical investigation of the olfactory organ in the salmon trout *Oncorhynchus masou* and chum salmon *O. keta. Journal of Ichthyology*, *45*, 185–190. doi:10.1093/chemse/bjp042

Kudo, H., Shinto, M., Sakurai, Y., & Kaeriyama, M. (2009). Morphometry of olfactory lamellae and olfactory receptor neurons during the life history of chum salmon (*Oncorhynchus keta*). *Chemical Senses*, *34*(7), 617–624. doi:10.1093/chemse/bjp042 PMID:19587025

Kudo, H., Ueda, H., Mochida, K., Adachi, S., Hara, A., Nagasawa, H., ... Yamauchi, K. (1999). Salmonid olfactory system-specific protein (N24) exhibits glutathione S-transferase class pi-like structure. *Journal of Neurochemistry, 72*(4), 1344–1352. doi:10.1046/j.1471-4159.1999.721344.x PMID:10098835

Menco, B. P. M. (1980). Qualitative and quantitative freeze-fracture studies on olfactory and nasal respiratory epithelia surfaces of frog, ox, rat, and dog. II. Cell apices, cilia, and microvilli. *Cell and Tissue Research, 211*, 5–29. doi:10.1007/BF00234393 PMID:6967758

Moran, D. T., Rowley, J. C. III, Jafek, B. W., & Lovell, M. A. (1982). The fine structure of the olfactory mucosa in man. *Journal of Neurocytology, 11*(5), 721–746. doi:10.1007/BF01153516 PMID:7143026

Morinishi, F., Shiga, T., Suzuki, N., & Ueda, H. (2007). Cloning and characterization of an odorant receptor in five Pacific salmon. *Comparative Biochemistry and Physiology Part B, 148*(3), 329–336. doi:10.1016/j.cbpb.2007.06.010 PMID:17652000

Murata, S., Takasaki, N., Saitoh, M., Tachida, H., & Okada, N. (1996). Details of retropositional genome dynamics that provide a rationale for a generic division: The distinct branching of all the Pacific salmon and trout (*Oncorhynchus*) from the Atlantic salmon and trout (*Salmo*). *Genetics, 142*, 915–926. doi:10.1006/gcen.2001.7646 PMID:8849897

Nevitt, G. A., Dittman, A. H., Quinn, T. P., & Moody, W. J. Jr. (1994). Evidence for a peripheral olfactory memory in imprinted salmon. *Proceedings of the National Academy of Sciences of the United States of America, 91*(10), 4288–4292. doi:10.1073/pnas.91.10.4288 PMID:7514302

Niimura, Y. (2009). Evolutionary dynamics of olfactory receptor genes in chordates: Interaction between environments and genomic contents. *Human Genomics, 4*(2), 107–118. doi:10.1186/1479-7364-4-2-107 PMID:20038498

Nordeng, H. A. (1971). Is the local orientation of anadromous fishes determined by pheromones? *Nature, 233*(5319), 411–413. doi:10.1038/233411a0 PMID:16063406

Nordeng, H. A. (1977). A pheromone hypothesis for homeward migration in anadromous salmonids. *Oikos, 28*(2/3), 155–159. doi:10.2307/3543965

Nosyk, O., Haseborg, E. T., Metzger, U., & Frimmel, F. H. (2008). A standardized pre-treatment method of biofilm flocs for fluorescence microscopic characterization. *Journal of Microbiological Methods*, *75*(3), 449–456. doi:10.1016/j.mimet.2008.07.024 PMID:18718852

Palstra, A. P., Fukaya, K., Chiba, H., Dirks, R. P., Planas, J. V., & Ueda, H. (2015). The olfactory transcriptome and progression of sexual maturation in homing chum salmon *Oncorhynchus keta*. *PLoS One*, *10*(9), e0137404. doi:10.1371/journal.pone.0137404 PMID:26397372

Sandahl, J. F., Miyasaka, G., Koide, N., & Ueda, H. (2005). Olfactory inhibition and recovery in chum salmon (Oncorhynchus keta) following copper exposure. *Canadian Journal of Fisheries and Aquatic Sciences*, *63*(8), 1840–1847. doi:10.1139/f06-074

Sato, K., Shoji, T., & Ueda, H. (2000). Olfactory discriminating ability of lacustrine sockeye and masou salmon in various freshwaters. *Zoological Science*, *17*, 313–317. PMID:18494584

Scholz, A. T., Horrall, R. M., & Hasler, A. D. (1976). Imprinting to chemical cues: The basis for home stream selection in salmon. *Science*, *192*(4245), 1247–1249. doi:10.1126cience.1273590 PMID:1273590

Shimizu, M., Kudo, H., Ueda, H., Hara, A., Shimazaki, M., & Yamauchi, K. (1993). Identification and immunological properties of an olfactory system-specific protein in kokanee salmon (*Oncorhynchus nerka*). *Zoological Science*, *10*, 287–294.

Shoji, T., Ueda, H., Ohgami, T., Sakamoto, T., Katsuragi, Y., Yamauchi, K., & Kurihara, K. (2000). Amino acids dissolved in stream water as possible homestream odorants for masu salmon. *Chemical Senses*, *25*(5), 533–540. doi:10.1093/chemse/25.5.533 PMID:11015325

Thomas, J. D. (1997). The role of dissolved organic matter, particularly free amino acids and humic substances, in freshwater ecosystems. *Freshwater Biology*, *38*(1), 1–36. doi:10.1046/j.1365-2427.1997.00206.x

Ueda, H. (2011). Physiological mechanisms of homing migration in Pacific salmon from behavioral to molecular biological approaches. *General and Comparative Endocrinology*, *170*(2), 222–232. doi:10.1016/j.ygcen.2010.02.003 PMID:20144612

Ueda, H. (2019). Sensory mechanisms of natal stream imprinting and homing in *Oncorhynchus* spp. *Journal of Fish Biology*, *95*(1), 293–33. doi:10.1111/jfb.13775 PMID:30101534

Ueda, H., Kudo, H., Shimizu, M., Mochida, K., Adachi, S., & Yamauchi, K. (1993). Immunological similarity between an olfactory system-specific protein and a testicular germ cell protein in kokanee salmon (*Oncorhynchus nerka*). *Zoological Science*, *10*, 685–690.

Ueda, H., Nakamura, S., Nakamura, T., Inada, K., Okubo, T., Furukawa, N., ... Watanabe, M. (2016). Involvement of hormones in olfactory imprinting and homing in chum salmon. *Scientific Reports*, *6*(1), 21102. doi:10.1038rep21102 PMID:26879952

Ueda, H., Shimizu, M., Kudo, H., Hara, A., Hiroi, O., Kaeriyama, M., ... Yamauchi, K. (1994). Species-specificity of an olfactory system-specific protein in various species of teleosts. *Fisheries Science*, *60*(2), 239–240. doi:10.2331/fishsci.60.239

Ueda, K. (1985). An electrophysiological approach to the olfactory recognition of homestream waters in chum salmon. *NOAA Technical Report NMFS*, *27*, 97–102.

Wickens, A., May, D., & Rand-Weaver, M. (2001). Molecular characterisation of a putative Atlantic salmon (*Salmo salar*) odorant receptor. *Comparative Biochemistry and Physiology Part B*, *129*(2-3), 653–660. doi:10.1016/S1096-4959(01)00364-5 PMID:11399502

Wisby, W. J., & Hasler, A. D. (1954). Effect of olfactory occlusion on migrating silver salmon (*Oncorhynchus kisutch*). *Journal of the Fisheries Research Board of Canada*, *11*(4), 472–478. doi:10.1139/f54-031

Yamamoto, M., & Ueda, K. (1977). Comparative morphology of fish olfactory epithelium-1. Salmoniformes. *Bulletin of Japanese Society of Scientific Fishers*, *43*(10), 1163–1174. doi:10.2331uisan.43.1163

Yamamoto, Y., Hino, H., & Ueda, H. (2010). Olfactory imprinting of amino acids in lacustrine sockeye salmon. *PLoS One*, *5*(1), e8633. doi:10.1371/journal.pone.0008633 PMID:20062811

Yamamoto, Y., Ishizawa, S., & Ueda, H. (2008). Effects of amino acid mixtures on upstream selective movement of four Pacific salmon. *Cybium*, *32*, 57–58.

Yamamoto, Y., Shibata, H., & Ueda, H. (2013). Olfactory homing of chum salmon to stable compositions of amino acids in natal stream water. *Zoological Science*, *30*(8), 607–612. doi:10.2108/zsj.30.607 PMID:23915152

Yamamoto, Y., & Ueda, H. (2009). Behavioral responses by migratory chum salmon to amino acids in natal stream water. *Zoological Science*, *26*(11), 778–782. doi:10.2108/zsj.26.778 PMID:19877838

Chapter 4
Memory in Salmon

ABSTRACT

It is unknown how salmon can imprint and retrieve information on their natal stream over a long period due to the lack of brain molecular markers for evaluating olfactory memory formation and retrieval. Memory in the brains of vertebrates is explained by the plasticity of the nervous system and the synaptic plasticity that promotes the ability of the chemical synapses to undergo changes in synaptic strength for long-term potentiation via the N-methyl-D-aspartate receptor, which has been identified a good molecular marker in the brain of salmon. This chapter describes the plasticity of the nervous system and synaptic plasticity, the involvement of the N-methyl-D-aspartate receptor in olfactory memory formation and retrieval in Pacific salmon, and functional magnetic resonance imaging of olfactory memory in lacustrine sockeye salmon.

DOI: 10.4018/978-1-7998-2054-3.ch004

INTRODUCTION

The brain of vertebrates is the center of the nervous system and collects information from outside (sensory system), determines the response to that information inside the body (motor system), and stores and retrieves the information when needed (memory). The core component of the nervous system is the neuron, which is an electrically excitable cell that processes and transmits information via electrochemical signaling. Memory is considered an information processing system, where sensory memory exists for less than one second, short-term (working) memory for several seconds to a minute, and long-term memory for an unlimited duration.

In ethology, imprinting is defined as phase-sensitive memory formation (learning) that occurs at a specific life stage of vertebrates. Filial imprinting was first studied by the English biologist Douglas Spalding using domestic chickens in the 19th century and was then rediscovered by German biologist Oskar Heinroth and studied intensively by his disciple Austrian biologist Konrad Lorenz, who shared the 1973 Nobel Prize in Physiology or Medicine with Nikolaas Tinbergen and Karl von Frisch. Lorenz discovered that graylag geese could imprint on the first moving stimulus observed between 13 and 16 hours after hatching, which was defined as the critical period.

Salmon imprinting on natal stream odors is closely related to filial imprinting in birds because salmon imprinting is considered to occur during juvenile downstream migration from their natal stream to the ocean, which should be the critical period. Moreover, salmon imprinting memory can remain over the long term until salmon produce offspring during adult homing from the ocean to their natal river several years later. Salmon imprinting should be defined as being characterized by a critical period as well as the long-term memory retention that is never lost.

Studies on hormones in salmon (Chapter 2) and olfaction in salmon (Chapter 3) have clearly indicated the hormonal control olfactory mechanisms of natal stream imprinting and homing in salmon. However, it is almost impossible to explain how hormones and olfaction can control imprinting and homing in salmon because of the lack of molecular markers that permit the evaluation of olfactory memory formation and retrieval in the brain of salmon.

Plasticity of Nervous System and Synaptic Plasticity

Memory in the brains of vertebrates is explained by the plasticity of the nervous system and synaptic plasticity. The plasticity of the nervous system is the structural and functional changes of the nervous system that occur in response to stimuli from outside environments. During the critical period of rapid development, active synaptogenesis, in which synaptic contact forms between the axon terminal of one neuron and the dendrite spine of another neuron, generates complex neural networks from new neural circuits or the exchange of neural circuits. Synaptic plasticity promotes the ability of chemical synapses to undergo a change in synaptic strength. As memory is thought to be encoded by modifications in synaptic strength, long-term potentiation (LTP), which is the persistent strengthening of synaptic activity that produces a long-lasting increase in signal transmission between two neurons, is widely considered one of the major cellular mechanisms that underlies memory (Kotaleski & Blackwell, 2010). LTP has been studied with a specific focus on the *N*-methyl-D-aspartate (NMDA) receptor, which induces LTP (Martin et al., 2000).

The NMDA receptor is a glutamate receptor channel subtype that mediates most of the fast-excitatory synaptic transmission in the central nervous system. In the case of normal low-frequency neurotransmission, the NMDA receptor is blocked by magnesium ions (Mg^{2+}). In contrast, in the case of critical high frequency neurotransmission, a substantial and rapid increase in the calcium ion concentration (Ca^{2+} flux) inside the postsynaptic dendrite spine via lifting of the magnesium blockade causes the activation of Ca/calmodulin-dependent-protein kinase II, protein kinase C, and FYN kinase. This kinase cascade series causes LTP induction (Figure 1). The NMDA receptor plays an important role in memory formation and retrieval in mammals (Park et al., 2013) and fish (Kinoshita et al., 2005; Nam et al., 2004; Cox et al., 2005; Gómez et al., 2006; Tzeng et al., 2007; Sison & Gerlai, 2011). The NMDA receptor is composed of two subunits: the essential NR1 subunit and the differentially expressed NR2A-D subunit (Shipton & Paulsen, 2013). The *nr1* gene of chum salmon has been cloned and characterized (Yu et al., 2014), and the effects of changes in salinity on *nr1* gene expression have been reported (Kim et al., 2015). However, there have been no reports addressing how NR1 might be involved in olfactory memory formation and retrieval in Pacific salmon.

Figure 1. Schematic illustration of normal low-frequency and critical high-frequency neurotransmission between the axon terminal and the dendrite spine. In the critical period, the Ca^{2+} flux through the NMDA receptor resulting from Mg^{2+} blockade lifting causes long-term potentiation (LTP) for the induction of synaptic plasticity.

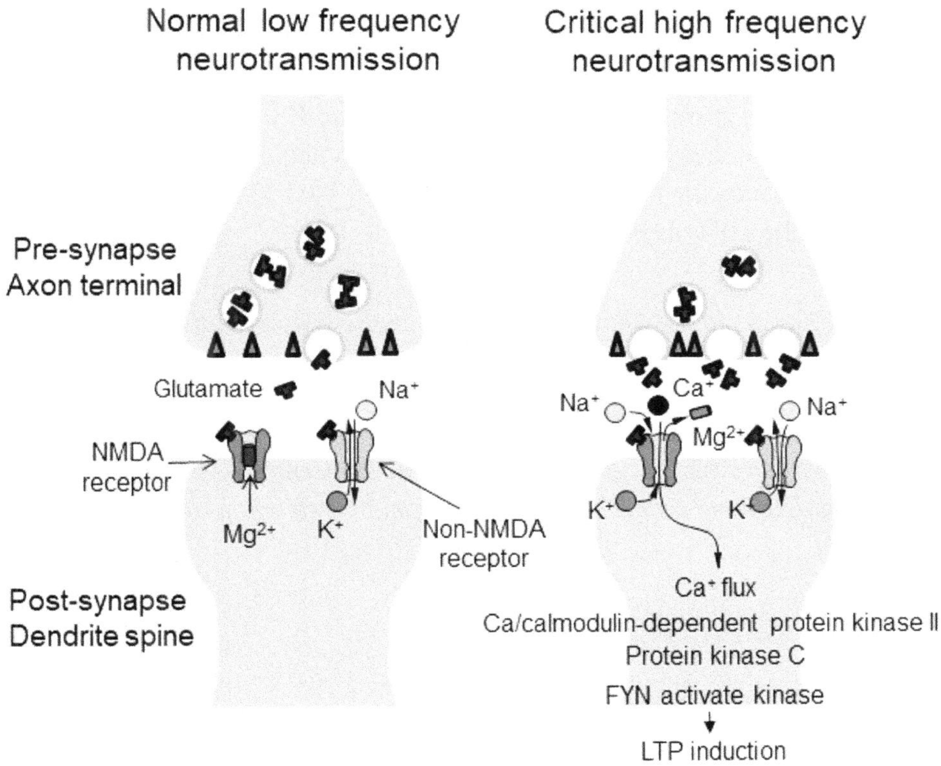

INVOLVEMENTS OF NR1 ON OLFACTORY MEMORY IN PACIFIC SALMON

Anadromous Chum Salmon

During the downstream migration of juvenile chum salmon from the Chitose Hatchery (Site **A**; Figure 5 in Chapter 2) to Ishikari Bay (Site **F**), the whole-brain mRNA expression level of *nr1* showed a significant initial peak that was correlated with larval brain development in the hatchery

(Pouwels, 1978); additionally, there was a gradual significant increase at the Chitose River in Kamaka Ward (Site **C**), and there was a second significant peak at the mouth of the Ishikari River (Site **E**) (Figure 2). Moreover, the results of *in situ* hybridization confirmed that, compared to the results in the hatchery (Site **A**), *nr1* mRNA localization increased prominently in the ventral TE in the Chitose River in Kamaka Ward (Site **C**) immediately after the fish were released into the river (Site **B**) (Ueda et al., 2016; Ueda, 2019). As shown in Figure 6 in Chapter 2, the *trha/b* gene expression showed significant peaks at Site **B** just after release from the Chitose Hatchery to the Chitose River, when acute environmental changes occurred, and *tshβ* gene expression and T4/T3 levels then increased suggesting the BPT axis activation.

*Figure 2. Changes in the whole-brain mRNA expression level of NMDA receptor essential subunit nr1 in juvenile chum salmon during downstream migration. Fish were sampled at the Chitose Hatchery (Site **A**) at four time points (January (**A1**), February (**A2**), March (**A3**), and April (**A4**)) and following release into the Chitose River from Site **B** to Site **F**. Data are presented as the means ± SEM (N=10-15). Different letters represent significant differences at p<0.05 level.*

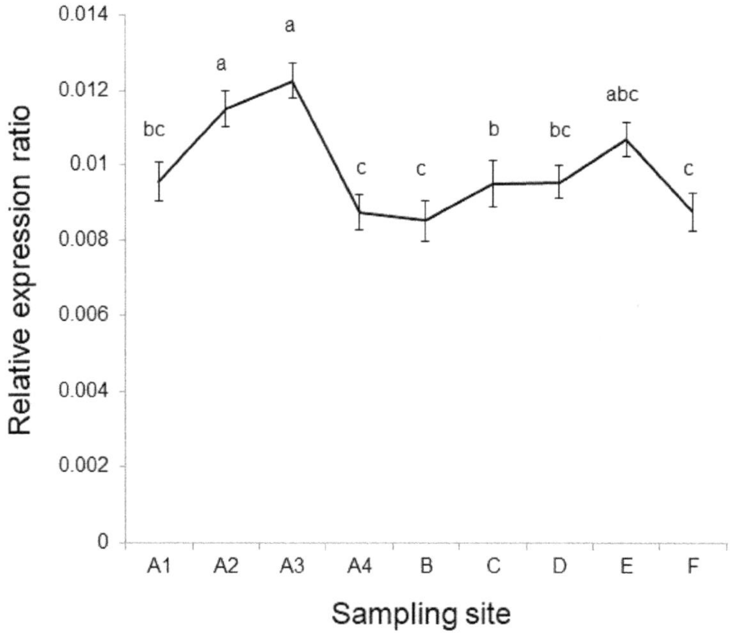

Figure 3. Effects of oral thyroxine (T4) administration on the gene expression ratios of the N-methyl-D-aspartate receptor essential subunit nr1 in the whole brain and T4 and triiodothyronine (T3) levels in the whole body of juvenile chum salmon. Data are presented as t

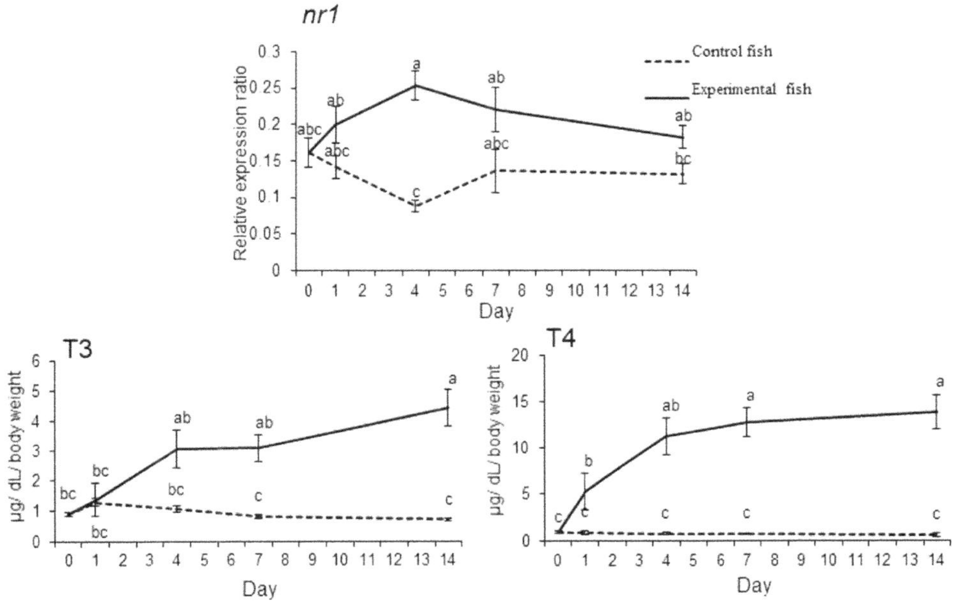

In experiments involving the oral administration of T4 (2 mg/g pellet) to juvenile chum salmon for 14 days, the levels of T4/T3 in the whole body increased for 14 days, and *nr1* gene expression in the whole brain increased significantly for 4 days (Ueda et al., 2016; Ueda, 2019). Interestingly, T4/T3 levels increased for 14 days, but no significant difference in *nr1* gene expression was observed; thus, these results revealed that the critical period of imprinting is limited to fewer than 7 days (Figure 3). These results, combined with the BPT hormone profiles during downstream migration in chum salmon (Figure 6 in Chapter 2) and the EOG response to waters from three streams in juvenile chum salmon during downstream migration (Figure 6 in Chapter 3), reveal that release from the hatchery into the river induces the activation of the BPT hormones and then stimulates the upregulation of NR1 related to olfactory memory formation. Further research should investigate how T4/T3 may activate the imprinting of natal stream odors in the olfactory epithelium.

Figure 4. Changes in the gene expression profiles of N-methyl-D-aspartate receptor essential subunit nr1 in the olfactory bulb and telencephalon of adult chum salmon during homing migration from the Bering Sea (Site G), Ishikari Bay (Site F), and the Indian waterwheel (Site J), to Chitose Hatchery (Site A). Data are presented as the means ± SEM (N=10). Different letters represent significant differences at p<0.05 level.

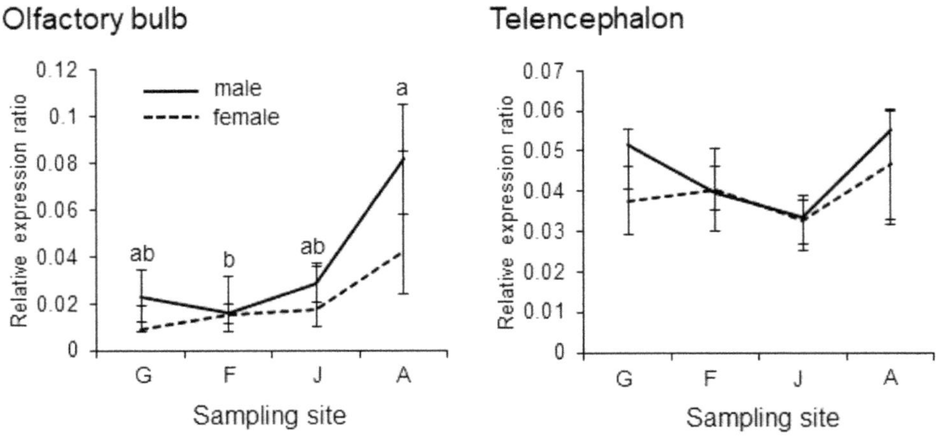

During the homing migration of adult chum salmon from the Bering Sea (Site **G**) to the Chitose Hatchery (Site **A**), the mRNA expression levels of *nr1* in the OB and TE appeared to increase, with some differences observed between males and females that might be related to sex-specific gonadal maturation (Figure 4). The increase in *nr1* gene expression in the OB showed that olfactory memory retrieval abilities increase during the upstream homing migration of adult chum salmon. Interestingly, *sgnrh-II* gene expression in the hypothalamus of males was also high in the Bering Sea (Ueda et al., 2016; Ueda, 2019). These results, combined with the *sgnrh-I/II* gene expression profiles in the OB and TE of adult chum salmon during homing migration from Site **G** to Site **A** (Figure 9 in Chapter 2) and the effects of GnRHa implantation on the EOG response to natal stream water in adult chum salmon collected from Site **F** (Figure 10 in Chapter 3), reveal that the initiation of homing migration in the Bering Sea induces the activation of the BPG hormones and then stimulates the upregulation of NR1 related to olfactory memory retrieval.

Lacustrine Sockeye Salmon

To investigate the involvement of the NMDA receptor in the determination of the homing duration in lacustrine sockeye salmon in Lake Shikotsu, the effect of NMDA receptor antagonists (APV and MK-801) on homing duration in late October was investigated (Ueda, 2016). Although there were no rescue experiments involving agonists, the homing duration of male lacustrine sockeye salmon was found to be significantly prolonged by both blockers (Figure 5A). These results suggest that NMDA receptors are deeply involved in the retrieval of imprinting memory in lacustrine sockeye salmon. Since MK-801 is known to impair learning performance in zebrafish (*Danio rerio*) (Sison & Gerlai, 2011), the effects of MK-801 injection (0.1 μg/g BW 4 times every other day for 2 weeks in 1-year-old fish) on EOG responses in lacustrine sockeye salmon that were exposed to 1 μM Pro and Glu for 2 weeks in 1-year-old fish were examined 2 years later in the spawning period (3-year-old fish). The EOG response to Pro and Glu in the MK-801-injected fish was significantly lower compared to that in the control fish (Figure 5B). These results indicate that the effects of NMDAR receptor-dependent olfactory memory formation are blocked for two years.

Since sGnRH neurons exist in the OB, TN, and POA of the brain, these sGnRH neurons are considered to play different roles during the homing migration of Pacific salmon. However, it is difficult to examine the functional differences in these sGnRH neurons *in vivo*. A previous report demonstrated that NMDA treatment could stimulate GnRH release in hypothalamic fragments in rats (López et al. 1992). The effects of *in vitro* NMDA treatment on sGnRH release were examined in brain slices obtained from lacustrine sockeye salmon (3-year-old) in the nonbreeding (June) and breeding periods (October) (Ueda, 2016). Each slice was stimulated with 5 mM NMDA in Mg^{2+}-free fish saline in samples collected from both sexes in June and October. In comparison with the fish sampled between June and October, the NMDA treatment significantly increased sGnRH release in the POA slices. In contrast, the NMDA treatment induced sGnRH release in the OB and TN slices, but no differences were observed between June and October (Figure 6). sGnRH neurons in the POA must be involved in gonadal maturation; however, these neurons in the OB and TN may present olfactory-related functions associated with olfactory-related homing migration. The sex specific roles of BPG hormones as well as the brain-specific roles of sGnRH in the sensory systems and thee central nervous systems should be examined in future studies.

These results confirm that in the case of juvenile salmon, at the initiation of juvenile salmon downstream migration, the environmental changes involved in the release from the hatchery into the river induce the activation of the BPT hormones, which then stimulate the upregulation of NR1 and induce the induction of LTP, ultimately enhancing olfactory memory formation related to natal stream-specific odors (Figure 7A). In the case of adult salmon, the sGnRH neurons in the POA must be involved in gonadal maturation, and those in the OB and TN may have olfactory-related functions that are related to olfactory-guided homing migration. The initiation of homing migration in adult salmon induces the activation of BPG hormones, which then stimulate the upregulation

*Figure 5. Effects of MNDA blockers (APV and MK-801) on the homing duration of male lacustrine sockeye salmon in Lake Shikotsu in October (A). EOG response to the test water (1 μM Pro and Glu) in MK-801-injected and control lacustrine sockeye salmon (B). Data are presented as the means ± SEM (A, n=10; B, n=3). Significant differences are indicated at *p<0.05 level.*

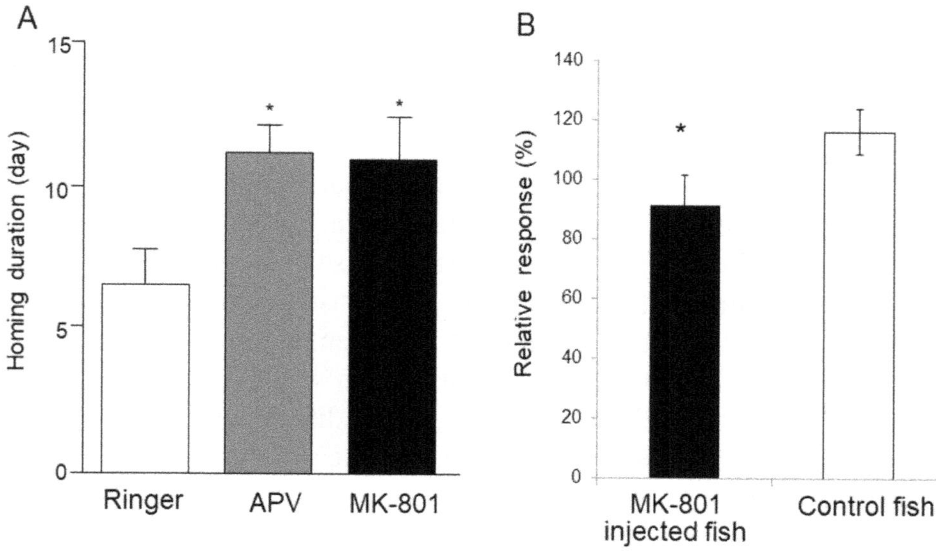

*Figure 6. Changes in the amount of sGnRH release from the olfactory bulb (OB), terminal nerve (TN), and preoptic area (POA) slices of the brain of lacustrine sockeye salmon in June and October. The release of sGnRH is significantly increased only in POA slices both in control and 5 mM MNDA in October. Data are presented as the means ± SEM (N=10). Significant differences are indicated at **p<0.01 levels.*

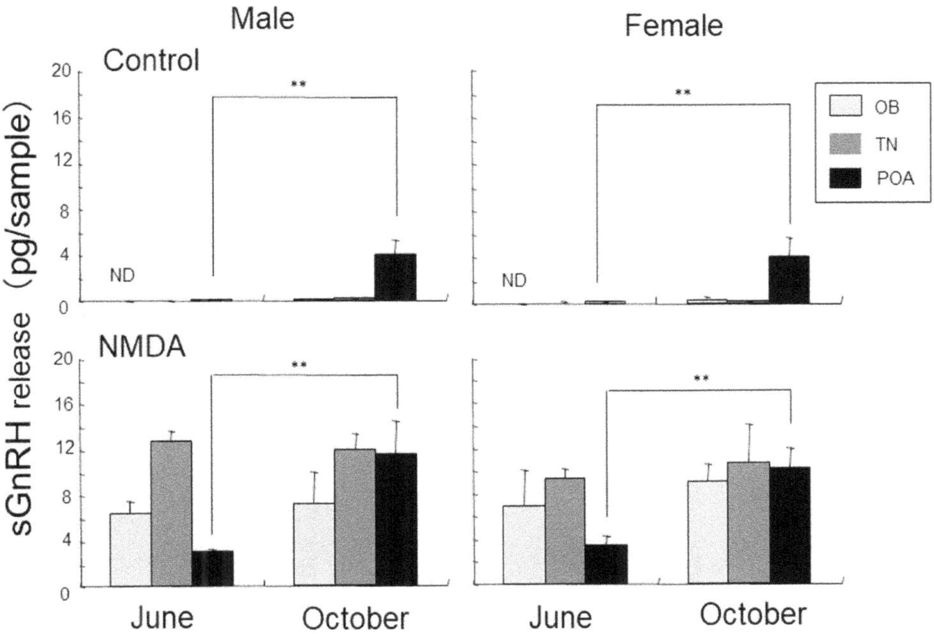

of NR1 and induce the induction of LTP, which subsequently improves the olfactory memory retrieval of natal stream-specific odors (Figure 7B). Further research should investigate how NR2A-D subunits are involved in olfactory memory formation and retrieval in Pacific salmon.

Functional Magnetic Resonance Imaging on Olfactory Memory

Functional magnetic resonance imaging (fMRI) is a noninvasive method that can measure neuronal activity via changes in cerebral blood flow and metabolism (Hyder et al., 2001).

Bandoh et al. (2011) applied blood oxygenation level-dependent (BOLD) fMRI to investigate natal stream odor information processing in the brain of lacustrine sockeye salmon. Strong responses to the natal stream water

Figure 7. Schematic illustrations of olfactory imprinting in juvenile salmon (A) and olfactory homing in adult salmon (B). The initiation of the downstream migration of juveniles induces the expression of the brain-pituitary-thyroid (BPT) hormones, which then induce the upregulation of the NMDA receptor, inducing long-term potentiation (LTP), which enhances olfactory memory formation related to natal stream-specific odors. The initiation of the homing migration of adult salmon induces the activation of the brain-pituitary-gonad (BPG) hormones, which then induce the upregulation of NMDA receptor, inducing LTP, which enhances olfactory memory retrieval related to natal stream-specific odors.

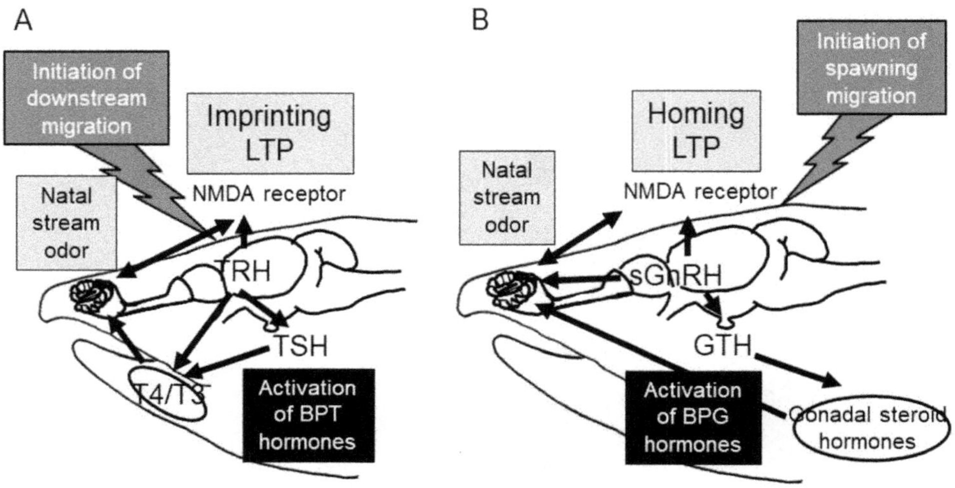

were mainly observed in the lateral area of the dorsal TE. Although the concentration of L-serine (1 mM) in the control water was 20,000-times higher than that of total amino acids in the natal stream water (47.5 nM), the BOLD signals resulting from the natal stream water were stronger than those induced by L-serine in the lateral area of dorsal TE (Figure 8). It was concluded that lacustrine sockeye salmon could process the odor information from the natal stream by integrating information in the lateral area of the dorsal TE.

Because of large differences between the development of the TE in teleosts via the eversion process and its development in higher vertebrates via the inversion process (Liem et al., 2001), it is difficult to compare the structural homology of the TE between teleosts and terrestrial vertebrates.

Figure 8. Time course of the BOLD fMRI signal changes induced by L-serine and natal stream water in the activation area of the olfactory bulb (A) and telencephalon (B: fontal area; C: middle area). These responses are averaged for six stimulus paradigms from the same fish. The locations of each area are indicated with boxes in the inserted MRI images. The bars indicate the stimulus period (3 min).

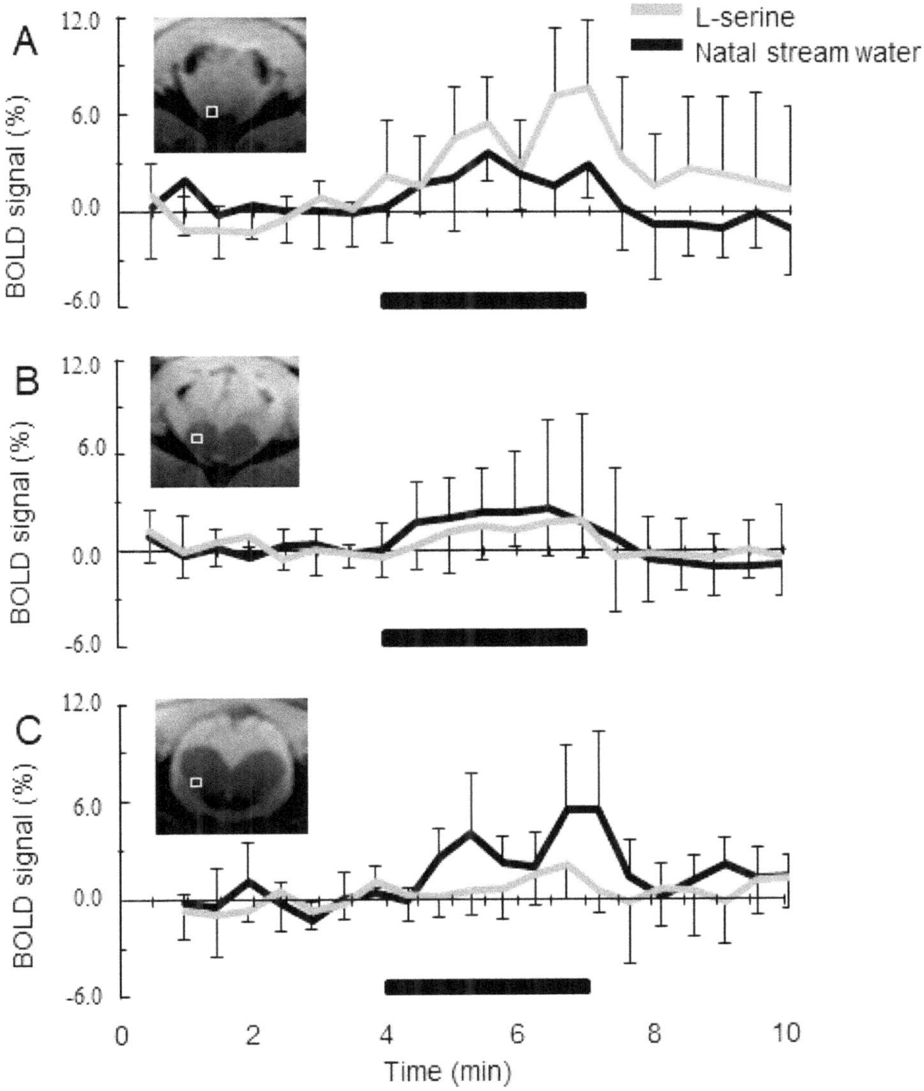

However, recent studies using a molecular marker in the medaka (*Oryzias latipes*) brain showed that the dorsal and ventral areas of TE were homologous to the pallium and subpallium in mammals, respectively (Kage et al., 2004; Alunni et al., 2004). Several studies have proposed that the dorsal TE in teleosts is homologous to the medial pallium (hippocampus) in terrestrial vertebrates (Wullimann & Mueller, 2004; Nikonov et al., 2005; Northcutt, 2006a, 2006b). Studies involving a partial lesion of the dorsal TE have reported that it plays a crucial role in the acquisition of spatial learning (Rodriguez et al., 2002; Saito & Watanabe, 2004). These results suggest that the dorsal TE has important functions in spatial memory formation and retrieval in lacustrine sockeye salmon.

CONCLUSION

The plasticity of the nervous system and synaptic plasticity are fundamental mechanisms for memory formation and retrieval in the brains of vertebrates. The plasticity of the nervous system results from the structural and functional changes of the nervous system that occur in response to stimuli from the outside environment. Synaptic plasticity promotes the ability of chemical synapses to change synaptic strength that is modified to encode memory formation. LTP is the persistent strengthening of synaptic activity that produces a long-lasting increase in the signal transmission between two neurons that is induced by the NMDA receptor. The NMDA receptor plays an important role in memory formation and retrieval and is composed of the essential NR1 subunit and the differentially expressed NR2A-D subunit. In juvenile salmon, the environmental changes associated with the release from the hatchery into the river at the initiation of downstream migration induce the activation of the BPT hormones, which then stimulate the upregulation of NR1 and induce the induction of LTP, ultimately enhancing olfactory memory formation related to natal stream-specific odors. In adult salmon, the sGnRH neurons in the POA must be involved in gonadal maturation, and those in the OB and TN may have olfactory-related functions that are related to olfactory-guided homing migration. The initiation of homing migration in adult salmon induces the activation of the BPG hormones, which then stimulate the upregulation of NR1 and induce the induction of LTP, which subsequently improves olfactory memory retrieval related to natal stream-specific odors. Further molecular biological studies should investigate how the NR2A-D subunits are involved in olfactory memory formation and retrieval in Pacific salmon. The examination of the effect of NMDA receptor antagonists (APV and MK-801) on the homing duration of lacustrine sockeye salmon in Lake Shikotsu revealed that homing duration was significantly prolonged. In addition, the effects of MK-801 injection on EOG responses in lacustrine sockeye salmon over 2 weeks in 1-year-old fish were examined 2 years later in the spawning period, and a significantly lower EOG response was observed in MK-801-injected fish. Bold fMRI was applied to investigate natal stream odor information processing in the brain of lacustrine sockeye salmon, and strong responses to the natal stream water were mainly observed in the lateral area of dorsal TE, suggesting that the dorsal TE has important functions in spatial memory formation and retrieval in lacustrine sockeye salmon.

REFERENCES

Alunni, A., Blin, M., Deschet, K., Bourrat, F., Vernier, P., & Rétaux, S. (2004). Cloning and developmental expression patterns of Dlx2, Lhx7, and Lhx9 in the medaka fish (*Oryzias latipes*). *Mechanisms of Development, 121*(7-8), 977–983. doi:10.1016/j.mod.2004.03.023 PMID:15210202

Bandoh, H., Kida, I., & Ueda, H. (2011). Olfactory responses to natal stream water in sockeye salmon by BOLD fMRI. *PLoS One, 6*(1), e16051. doi:10.1371/journal.pone.0016051 PMID:21264223

Cox, J. A., Kucenas, S., & Voigt, M. M. (2005). Molecular characterization and embryonic expression of the family of N-methyl-D-aspartate receptor subunit genes in zebrafish. *Developmental Dynamics, 234*(3), 756–766. doi:10.1002/dvdy.20532 PMID:16123982

Gómez, Y., Vargas, J. P., Portavella, M., & López, J. C. (2006). Spatial learning and goldfish telencephalon NMDA receptors. *Neurobiology of Learning and Memory, 85*(3), 252–262. doi:10.1016/j.nlm.2005.11.006 PMID:16464619

Hyder, F., Kida, I., Behar, K. L., Kennan, R. P., Maciejewski, P. K., & Rothman, D. L. (2001). Quantitative functional imaging of the brain: Towards mapping neuronal activity by BOLD fMRI. *NMR in Biomedicine, 14*(7-8), 413–431. doi:10.1002/nbm.733 PMID:11746934

Kage, T., Takeda, H., Yasuda, T., Maruyama, K., Yamamoto, N., Yoshimoto, M., ... Ishikawa, Y. (2004). Morphogenesis and regionalization of the medaka embryonic brain. *The Journal of Comparative Neurology, 476*(3), 219–239. doi:10.1002/cne.20219 PMID:15269967

Kim, N. N., Choi, Y. J., Lim, S. G., Jeong, M., Jin, D. H., & Choi, C. Y. (2015). Effect of salinity changes on olfactory memory-related genes and hormones in adult chum salmon *Oncorhynchus keta. Comparative Biochemistry and Physiology Part A, 187*, 40–47. doi:10.1016/j.cbpa.2015.04.011 PMID:25933936

Kinoshita, M., Fukaya, M., Tojima, T., Kojima, S., Ando, H., Watanabe, M., ... Ito, E. (2005). Retinotectal transmission in the optic tectum of rainbow trout. *The Journal of Comparative Neurology, 484*(2), 249–259. doi:10.1002/cne.20473 PMID:15736228

Kotaleski, J. H., & Blackwell, K. T. (2010). Modelling the molecular mechanisms of synaptic plasticity using systems biology approaches. *National Review, 11*(4), 239–251. doi:10.1038/nrn2807 PMID:20300102

Liem, K. F., Bemis, W. E., Walker, W. F., & Grande, L. (2001). *Functional Vertebrate Anatomy*. Philadelphia: Harcourt College Publishers.

López, F. J., Donoso, A. O., & Negro-Vilar, A. (1992). Endogenous excitatory amino acids and glutamate receptor subtypes involved in the control of hypothalamic luteinizing hormone-releasing hormone secretion. *Endocrinology, 130*, 1986–1992. doi:10.1210/endo.130.4.1312433 PMID:1312433

Martin, S. J., Grimwood, P. D., & Morris, R. G. (2000). Synaptic plasticity and memory: An evaluation of the hypothesis. *Annual Review of Neuroscience, 23*(1), 649–711. doi:10.1146/annurev.neuro.23.1.649 PMID:10845078

Nam, R. H., Kim, W., & Lee, C. J. (2004). NMDA receptor-dependent long-term potentiation in the telencephalon of the zebrafish. *Neuroscience Letters, 370*(2-3), 248–251. doi:10.1016/j.neulet.2004.08.037 PMID:15488332

Nikonov, A. A., Finger, T. E., & Caprio, J. (2005). Beyond the olfactory bulb: An odotopic map in forebrain. *Proceedings of the National Academy of Sciences of the United States of America, 102*(51), 18688–18693. doi:10.1073/pnas.0505241102 PMID:16339016

Northcutt, R. G. (2006a). Connection of the lateral and medial divisions of the goldfish telencephalic pallium. *The Journal of Comparative Neurology, 494*(6), 903–943. doi:10.1002/cne.20853 PMID:16385483

Northcutt, R. G. (2006b). Forebrain evolution in bony fishes. *Brain Research Bulletin, 75*(2-4), 191–205. doi:10.1016/j.brainresbull.2007.10.058 PMID:18331871

Park, P., Volianskis, A., Sanderson, T. M., Bortolotto, Z. A., Jane, D. E., Zhuo, M., ... Collingridge, G. L. (2013). NMDA receptor-dependent long-term potentiation comprises a family of temporally overlapping forms of synaptic plasticity that are induced by different pattern of stimulation. *Philosophical Transactions of the Royal Society of London. Series B, Biological Sciences, 369*(1633), 20130131. doi:10.1098/rstb.2013.0131 PMID:24298134

Pouwels, E. (1978). On the development of the cerebellum of the trout, *Salmo gairdneri*. *Anatomy and Embryology*, *152*(3), 291–308. doi:10.1007/BF00350526 PMID:655434

Rodriguez, F., López, J. C., Vargas, J. P., Gomez, Y., Broglio, C., & Salas, C. (2002). Conservation of spatial memory function in the pallial forebrain of reptiles and ray-finned fishes. *The Journal of Neuroscience*, *22*(7), 2894–2903. doi:10.1523/JNEUROSCI.22-07-02894.2002 PMID:11923454

Saito, K., & Watanabe, S. (2004). Spatial learning deficits after dorsomedial telencephalon lesions in the goldfish. *Neuroreport*, *15*, 2695–2699. PMID:15597037

Shipton, O. A., & Paulsen, O. (2013). GluN2A and GluN2B subunit-containing NMDA receptors in hippocampal plasticity. *Philosophical Transactions of the Royal Society of London. Series B, Biological Sciences*, *369*(1633), 20130163. doi:10.1098/rstb.2013.0163 PMID:24298164

Sison, M., & Gerlai, R. (2011). Behavioral performance altering effects of MK-801 in zebrafish (*Danio rerio*). *Behavioural Brain Research*, *220*(2), 331–337. doi:10.1016/j.bbr.2011.02.019 PMID:21333690

Tzeng, D. W., Lin, M. H., Chen, B. Y., Chen, Y. C., Chang, Y. C., & Chow, W. Y. (2007). Molecular and functional studies of tilapia (*Oreochromis mossambicus*) NMDA receptor NR1 subunits. *Comparative Biochemistry and Physiology. B, Comparative Biochemistry*, *146*(3), 402–411. doi:10.1016/j.cbpb.2006.11.026 PMID:17258920

Ueda, H. (2016). Physiological mechanisms of imprinting and homing migration of Pacific salmon. *Aqua-BioScience Monographs*, *9*(1), 1–27. doi:10.5047/absm.2016.00901.0001

Ueda, H. (2019). Sensory mechanisms of natal stream imprinting and homing in *Oncorhynchus* spp. *Journal of Fish Biology*, *95*(1), 293–33. doi:10.1111/jfb.13775 PMID:30101534

Ueda, H., Nakamura, S., Nakamura, T., Inada, K., Okubo, T., Furukawa, N., ... Watanabe, M. (2016). Involvement of hormones in olfactory imprinting and homing in chum salmon. *Scientific Reports*, *6*(1), 21102. doi:10.1038rep21102 PMID:26879952

Wullimann, M. F., & Mueller, T. (2004). Teleostean and mammalian forebrains contrasted: Evidence from genes to behavior. *The Journal of Comparative Neurology*, *475*(2), 143–162. doi:10.1002/cne.20183 PMID:15211457

Yu, J. N., Ham, S. H., Lee, S. I., Jin, H. J., Ueda, H., & Jin, D. H. (2014). Cloning and characterization of the N-methyl-D-aspartate receptor subunit NR1 gene from chum salmon, *Oncorhynchus keta* (Walbaum, 1792). *SpringerPlus*, *3*(1), 9. doi:10.1186/2193-1801-3-9 PMID:24422186

Chapter 5
Behavior in Salmon

ABSTRACT

The recent rapid development of biotelemetry technologies has made it possible to continuously observe the underwater behavior of salmon in open water. Homing migratory behaviors were studied using anadromous chum salmon from the Bering Sea to Hokkaido and lacustrine sockeye salmon and masu salmon in Lake Toya. Biotelemetry results on the migratory behavior of adult chum salmon in a reconstructed reach of the Shibetsu River; the investigation of cardiac arrest during gamete release in chum salmon; the comparison of the swimming ability and upstream-migration behavior of chum salmon and masu salmon in Hokkaido, Japan; and the analysis of site fidelity and habitat use in Formosan landlocked salmon during the typhoon season in the Chichiawan stream, Taiwan were also performed. This chapter describes the homing migration of anadromous chum salmon from the Bering Sea to Hokkaido, Japan; the homing migration of lacustrine sockeye salmon and masu salmon in Lake Toya, Hokkaido, Japan; and biotelemetry research on various behaviors in salmon.

DOI: 10.4018/978-1-7998-2054-3.ch005

INTRODUCTION

The recent rapid development of biotelemetry technologies, such as ultrasonic and radio telemetry, data logging, and pop-up satellite telemetry, has made it possible to continuously observe the underwater behavior of salmon in open water (Cooke et al., 2004; Ueda, 2004; Hussey et al., 2015). In particular, ultrasonic transmitters that emit pulsed signals have been useful for investigating the migratory behavior of salmon in coastal seas (Quinn & Groot, 1984; Quinn et al., 1989) and the central Bering Sea (Ogura & Ishida, 1994). In addition, some ultrasonic tracking experiments have been conducted with sensory ablation treatments. The behavior of anosmic Atlantic salmon was studied using ultrasonic tracking in a fjord and demonstrated the olfactory discrimination of fine-scale hydrographic features (Døving et al., 1985). The effects of visual and olfactory ablation on the swimming behavior of adult chum salmon were examined using ultrasonic tracking along the Okhotsk coast of Hokkaido, Japan, and this study reported vertical and horizontal zigzag movements in both types of ablated fish (Yano & Nakamura, 1992). Ultrasonic telemetry techniques were developed to investigate the magnetic sense of maturing chum salmon in the western North Pacific Ocean, and there were no significant differences in horizontal and vertical movements or in swimming speeds observed

Figure 1. Seasonal migration model of Japanese chum salmon estimated by genetic stock identification. Modified from Urawa (2004).

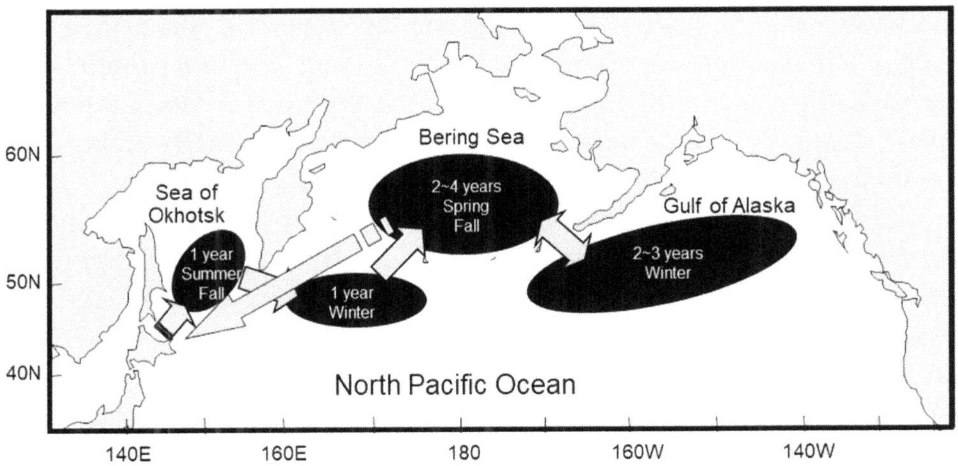

between the magnetically disturbed fish and the control fish (Yano et al., 1996). However, magnetoreceptor cells have been identified in the nasal cavity of rainbow trout (Walker et al., 1997), and empirical evidence of geomagnetic imprinting has been reported for Chinook salmon (Brancis & Anderson, 2012), pink salmon and sockeye salmon (Putman et al., 2013). Recently, the Earth's magnetic forces were studied during chum salmon homing migration from the Bering Sea to Japan using magnetic archival tags that can record the magnetic force; the reported results demonstrated that the homing migration route approximately followed the isoline of magnetic intensity (Azumaya et al., 2016). Recently, magnetoreception in fish was reviewed by Formicki et al. (2019).

Homing Migration of Anadromous Chum Salmon from the Bering Sea to Japan

A seasonal migration model for Japanese chum salmon in the Pacific Ocean has been proposed on the basis of genetic stock identification techniques, which suggested that homing migration from the Bering Sea to Japan is initiated in summer, mainly among 4-year-old fish (Figure 1) (Urawa, 2004). Tanaka et al. (2005) examined chum salmon homing behavior from the Bering Sea to Hokkaido, Japan. Japanese chum salmon were collected by using longlines in June 2000 in the central Bering Sea (56°30' N, 179°00' E), and the fish were identified as having a Japanese origin based on scale analysis; specifically, this was possible because most of the Japanese chum salmon were released from hatcheries, and the width of their scale rings during the fry stage was wider than that seen in wild salmon from other countries. In total, 27 chum salmon were released with an external propeller data logger that recorded swimming speed, depth, and temperature, which was attached externally through the dorsal musculature of the fish anterior to the dorsal fin.

Only one logger was recovered from a set net in September along the eastern coast of Hokkaido, Japan (43°20' N, 145°46' E) (Figure 2). The first recorded swimming profiles of chum salmon homing from the Bering Sea to Hokkaido were collected over a period of 67 days, and these profiles covered a linear distance of 2,750 km. The salmon travelled with an average swimming speed, depth, and temperature of 62 ± 12 (mean

Figure 2. Map indicating the release and recovery sites of a chum salmon carrying a propel data logger.

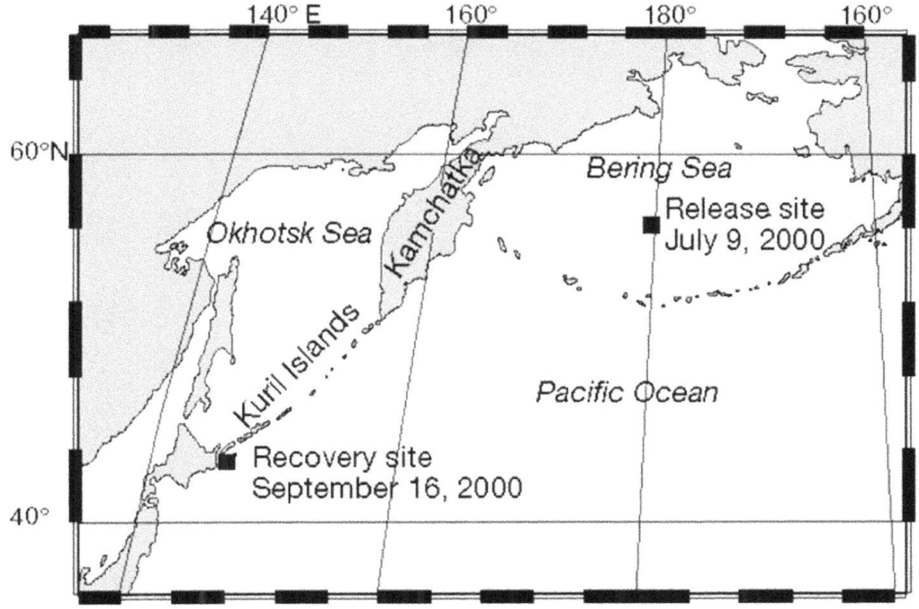

Figure 3. Depth, ambient temperature, and swimming speed of a chum salmon during open-ocean homing migration recorded by a propel data logger.

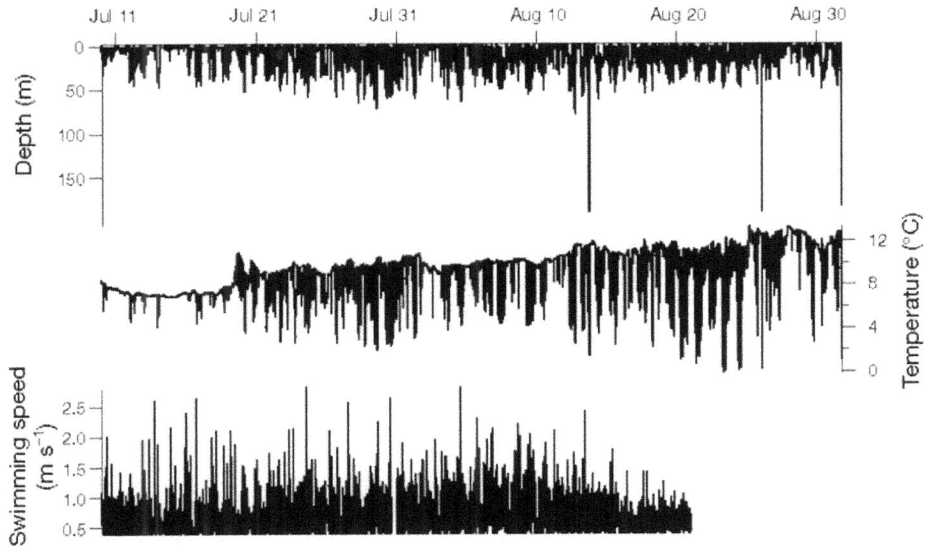

Figure 4. Hourly mean swimming depth and hourly mean swimming speed of a chum salmon during open-ocean homing migration.

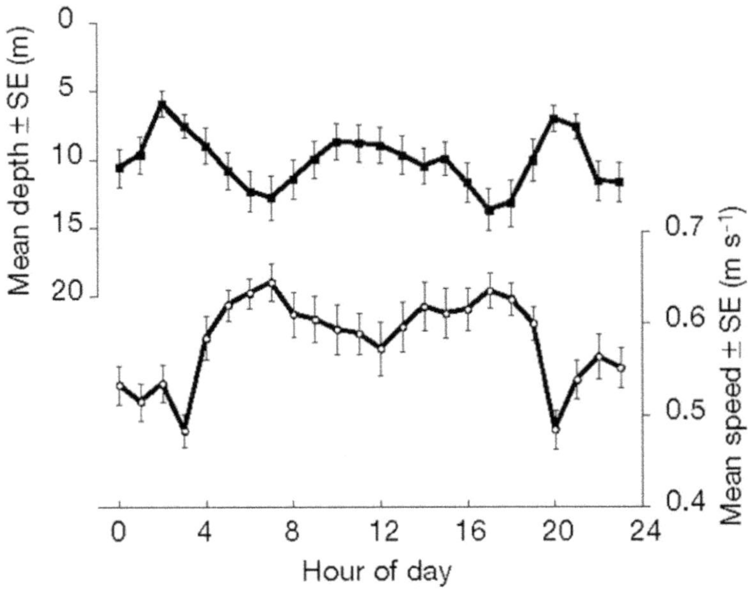

Figure 5. A 7-hour excerpt of the time-depth, temperature, and swimming speed data of a chum salmon showing a typical pattern of repeated v-shaped vertical movements (diving bouts) in the vicinity of the thermocline.

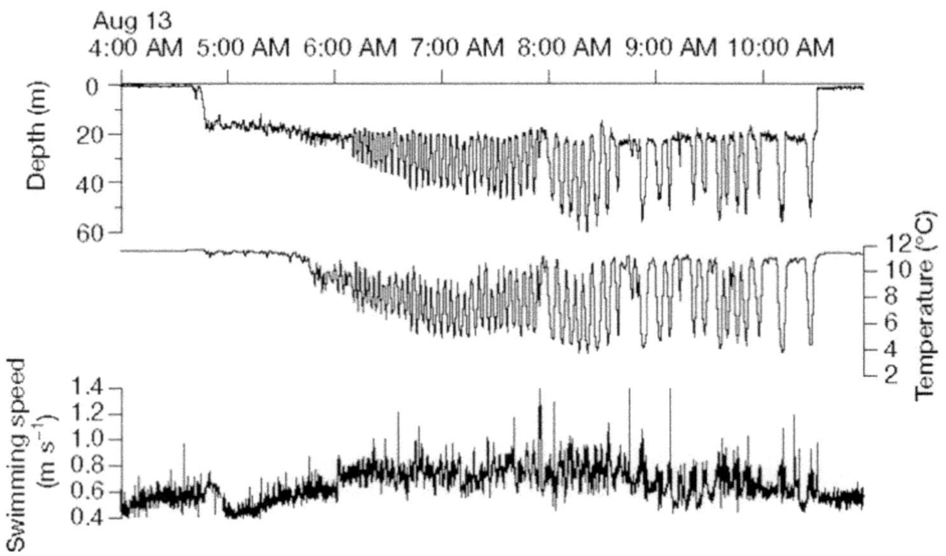

± SD) cm/sec, 10.4 ± 14.7 m, and 9.2 ± 0.2°C, respectively (Figure 3). Both swimming speed and depth trimodally peaked around dawn and dusk, with an additional small peak around midnight (Figure 4). The fish showed sequential upward and downward movements near the thermocline during twilight and daytime (Figure 5). These diurnal migratory patterns suggested that homing chum salmon allocated a proportion of their time to foraging and that their foraging strategy differed between daytime and nighttime. These results indicated that homing chum salmon exhibit homeward-bound navigation abilities from the central Bering Sea to Japan. Along the coast of Hokkaido and northern mainland of Japan, the swimming depth of chum salmon was recorded using archival tags; the results indicated that swimming depth was related to moonlight during homeward migrations (Hasegawa, 2012).

Homing Migration of Lacustrine Sockeye Salmon and Masu Salmon in Lake Toya, Hokkaido, Japan

It is difficult to implement physiologically controlled and manipulated experiments using anadromous salmon as the fish migrate from seawater in coastal seas to freshwater in their natal streams once they reach gonadal maturation. On the other hand, lacustrine salmon populations offer a good model system for studying homing behaviors during migration from open waters to natal areas for reproduction using sensory ablation treatments. The homing migrations of mature lacustrine sockeye salmon (Ueda et al., 1998) and lacustrine masu salmon (Ueda et al., 2000) whose sensory cues were impaired by ablation were compared, and the fish were tracked from the center of Lake Toya in Hokkaido, Japan, which is a large caldera lake (surface area 71 km^2; average and maximum depths of 116 m and 179 m, respectively), to their natal areas using an ultrasonic tracking system. A mature sockeye salmon male whose magnetic cues were interfered with by a strong NdFe magnetic ring that was attached to its head returned straight to the natal area after 1 hour of random movement. A mature sockeye salmon male whose visual cues were blocked by the injection of carbon tonner and corn oil and whose magnetic cues were altered with a magnetic ring traveled in a direction opposite from its natal area (Figure 6A). In contrast, one mature control masu salmon male traveled constantly along the coast and stopped at the mouth of stream, whereas a blinded mature masu salmon female that was released traveled randomly away from the coast, and a mature masu salmon male whose olfactory cues were blocked by Vaseline traveled first along the coast and eventually traveled towards the open water (Figure 6B).

The comparison between the straight movement of sockeye salmon and the coastal movement of masu salmon is quite informative. It has been reported that sockeye salmon are widely distributed in the North Pacific Ocean, while masu salmon are narrowly distributed in the western North Pacific Ocean (Kaeriyama & Ueda, 1998). These data suggest some evolutionary aspects of successful homing migration in Pacific salmon, where the narrowly distributed masu salmon require only coastal recognition ability for successful migration using olfactory and visual cues, while the widely distributed sockeye salmon require the development of open-water orientation ability to successfully return to their natal streams with high precision using visual and magnetic cues (Ueda, 2014). The roles of each sensory system should be further investigated, including how each sensory system controls coastal recognition and open-water orientation.

The spawning migration of lacustrine masu salmon with attached swimming depth and temperature data loggers was monitored in Lake Toy (Ueda, 2016). Clear diurnal movements were observed after they first encountered the mouth of the river at the beginning of the spawning season; this change was detected on the basis of the difference in the ambient temperature between the mouth of the river (14-20° C) and the surface of the lake (20-22° C) (Figure 7A). After this encounter, the fish swam vertically around the thermocline depth during the day and remained near the water surface at night, showing clear diurnal movements that could represent thermoregulation behavior (Figure 7B). These movements gradually disappeared towards the peak of the spawning season when they migrated upstream to the spawning ground (Figure 7C). These behavioral changes observed in masu salmon during the spawning season showed that masu salmon were able to calculate the day length with an innate biological clock; furthermore, this circadian rhythm after encountering cool temperatures at the river mouth was useful for calculating the timing of upstream migration to spawning sites.

Figure 6. Tracks of four mature male lacustrine sockeye salmon (A) and three mature lacustrine masu salmon (B) in Lake Toya during the spawning season determined using an ultrasonic tracking system. Arrowheads indicate the release point of each fish.

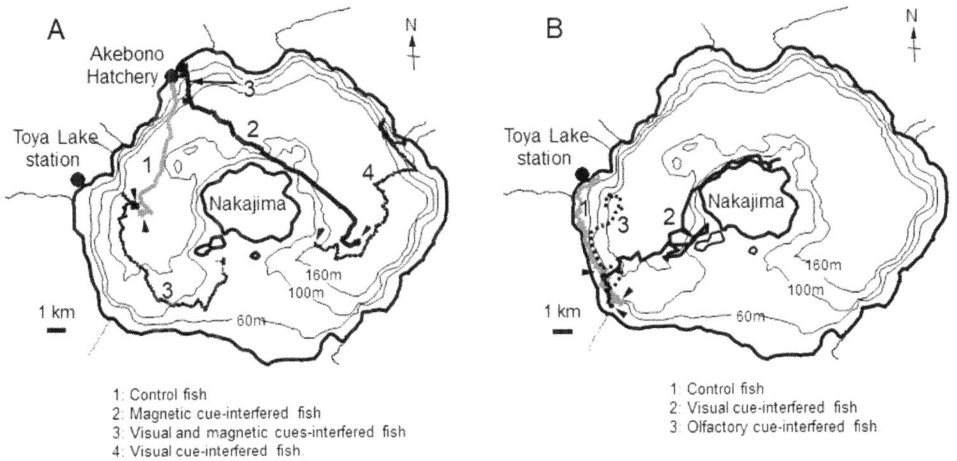

Figure 7. Swimming depth and ambient temperature of a masu salmon in Lake Toya during the first swimming period encountering the mouth of the stream (A), after swimming in the stream water (B) and just prior to upstream migration (C) recorded by a depth and temperature data logger.

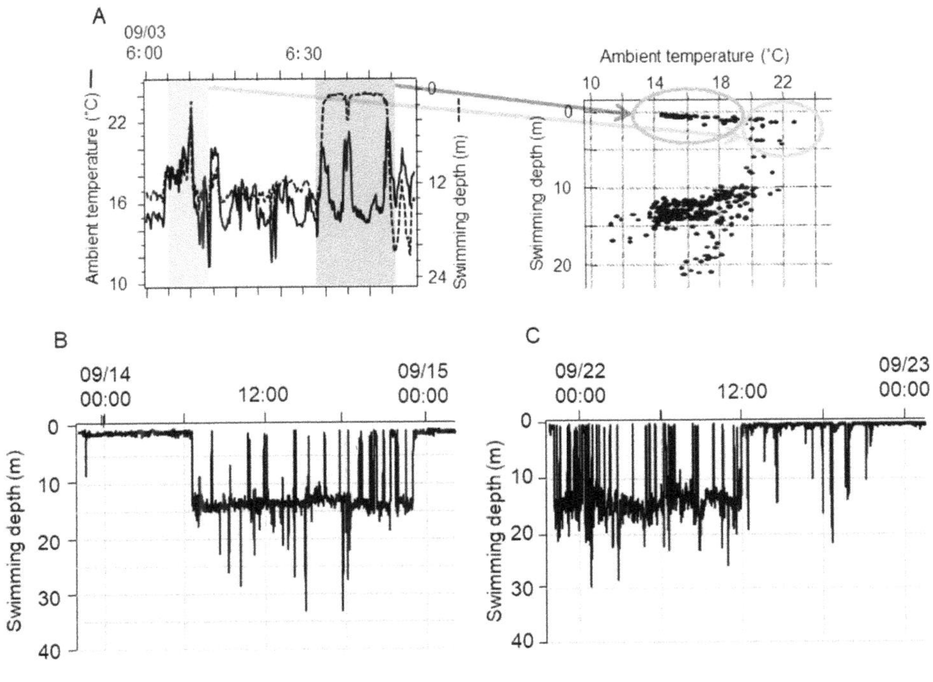

To accurately home in open water, salmon must recognize exact locations (map) and compass directions (orientation) during migration, and they must have an internal biological clock to calculate the duration of migration from the feeding ground to the spawning ground. New research approaches that reveal how salmon can navigate in open water using compasses, maps, and biological clocks will be necessary to identify which sensory systems are involved (and how) in the oceanic imprinting and homing migration of Pacific salmon.

BIOTELEMETRY STUDIES ON VARIOUS BEHAVIORS IN SALMON

Migratory Behavior of Adult Chum Salmon in a Reconstructed Reach of the Shibetsu River, Hokkaido, Japan

During 2004–2006, 30 adult chum salmon were tagged with electromyogram (EMG) transmitters and depth/temperature data loggers and tracked along both channelized and reconstructed segments of the Shibetsu River, Japan,

Figure 8. Holding time (hour) per individual chum salmon from 2004 to 2006 for the channelized segment (solid bars) and the reconstructed segment (open bar).

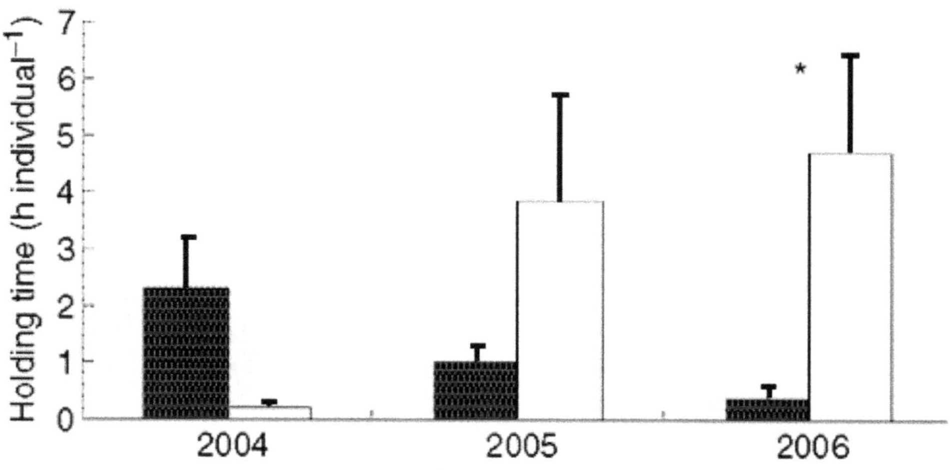

to assess the effects of river remeandering construction on upstream migration (Makiguchi et al., 2008). The holding time of chum salmon in the reconstructed segment was longer than that in the channelized segment (Figure 8). Furthermore, a greater diversity of physical conditions in water velocity was observed in the reconstructed segment in 2005. Swimming speeds decreased during holding and often exceeded the critical swimming speed (*Ucrit*). It appears that the physical characteristics developed in the reconstructed segment may be suitable for the holding of chum salmon. Furthermore, a swimming speed exceeding *Ucrit* prior to holding may be exhausting, so holding behavior may be important for muscle recovery as well as for efficient swimming to reduce energetic cost during upstream migration.

Cardiac Arrest During Gamete Release in Chum Salmon

Cardiac arrest caused by startling stimuli such as visual and vibration stimuli has been reported in some animals and could be considered an extraordinary case of bradycardia and defined as reversible missed heartbeats (Cuadras, 1980; Laming & Austin, 1981; Ide & Hoffmann, 2002; King & Adamo, 2006). The variability of the heart rate is established by the balance between the ANS (cholinergic vagus inhibition) and the excitatory adrenergic stimulation of neural and hormonal action in teleosts. However, cardiac arrest and its nervous regulatory mechanism remain poorly understood. By using electrocardiogram (ECG) data loggers, it was found that cardiac arrest occurs in chum salmon at the moment of gamete release for 7.4 s in females and for 5.2 s in males (Makiguchi et al., 2009a). The increase in heart rate during spawning behavior relative to the background rate during the resting period suggests that cardiac arrest is a characteristic physiological phenomenon of the extraordinarily high heart rate during spawning behavior. The ECG morphological analysis showed a peak and a tall T-wave adjacent to the cardiac arrest, indicating an increase in potassium permeability in cardiac muscle cells, which would function to retard the cardiac action potential. Pharmacological studies showed that cardiac arrest was abolished by injection of atropine, a muscarinic receptor antagonist, revealing that the observed cardiac arrest is a reflex response of the parasympathetic nervous system, although the injection of sotalol, a β-adrenergic antagonist, did not affect cardiac arrest (Figure 9). We conclude that cardiac arrest during gamete release during spawning release in spawning chum salmon is a physiological reflex response controlled by the parasympathetic nervous system. This cardiac arrest represents a response to the gaping behavior that occurs at the moment of gamete release.

Figure 9. Effects of pharmacological injections on R-R interval, cardiac arrest, and ECG morphology in female chum salmon. (A) Tachogram of 700 consecutive heart beats immediately after egg release in fish injected with atropine (upper line, N=3), sotalol (middle line, n=4), and Salmon Ringer as a sham control (lower line, n=3). (B) ECG at the moment of egg release in fish injected with atropine, sotalol, and Salmon Ringer. Vertical arrows indicate the start and end points of gaping. At the moment of egg release, peaked and tall T-waves occur in fish injected with sotalol and Salmon Ringer, but not in fish injected with atropine.

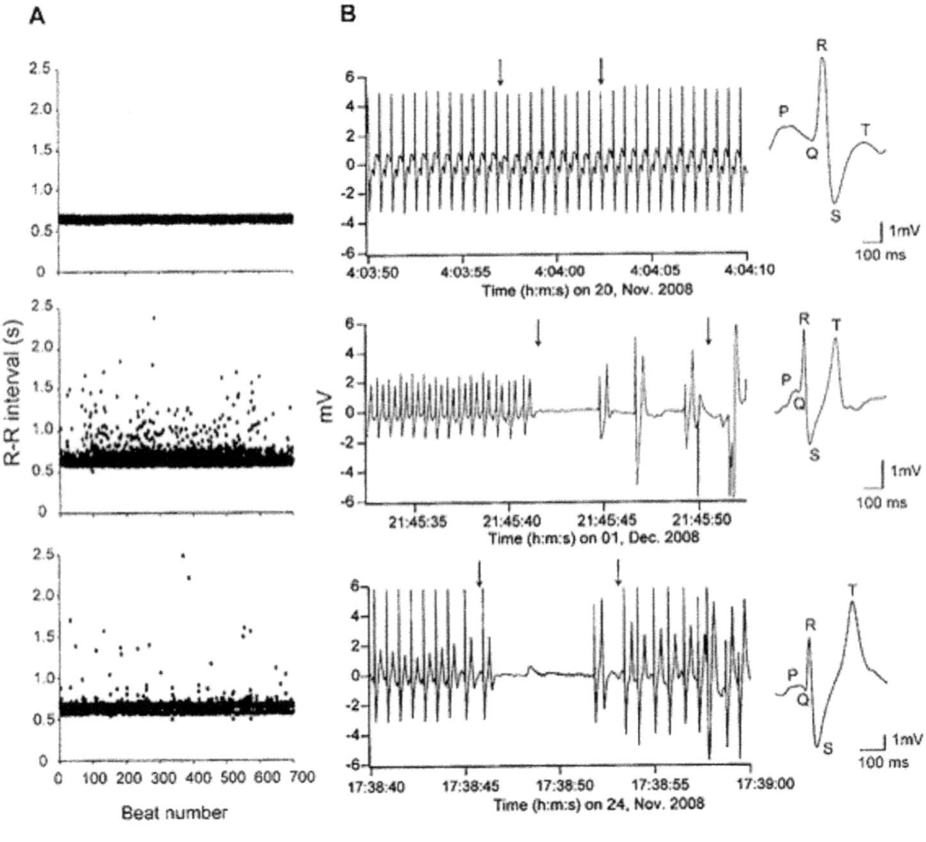

Site Fidelity and Habitat use of Formosan Landlocked Salmon During Typhoon Season in the Chichiawan Stream, Taiwan

The Formosan landlocked salmon is the most southerly native salmonid in the Northern Hemisphere, and it is only distributed in the Chichiawan Stream basin and Hsuehshan Creeks located in the Shei-Pa National Park in central Taiwan (Lin et al., 1990). The International Union for Conservation of Nature (IUCN) listed this species as a critically endangered species in the Red Data Book in 1996. Horizontal positioning, as an indicator of the movements and habitat use of adult Formosan landlocked salmon, was

Figure 10. Movement distance of six (A) and four tagged fish (B) during tracking periods. Symbol shapes denote each individual. Distance is toward upstream. Gray line shows water level measured in tracking section. Vertical arrow shows floods caused by typhoon Sepat on 18 August 2007.

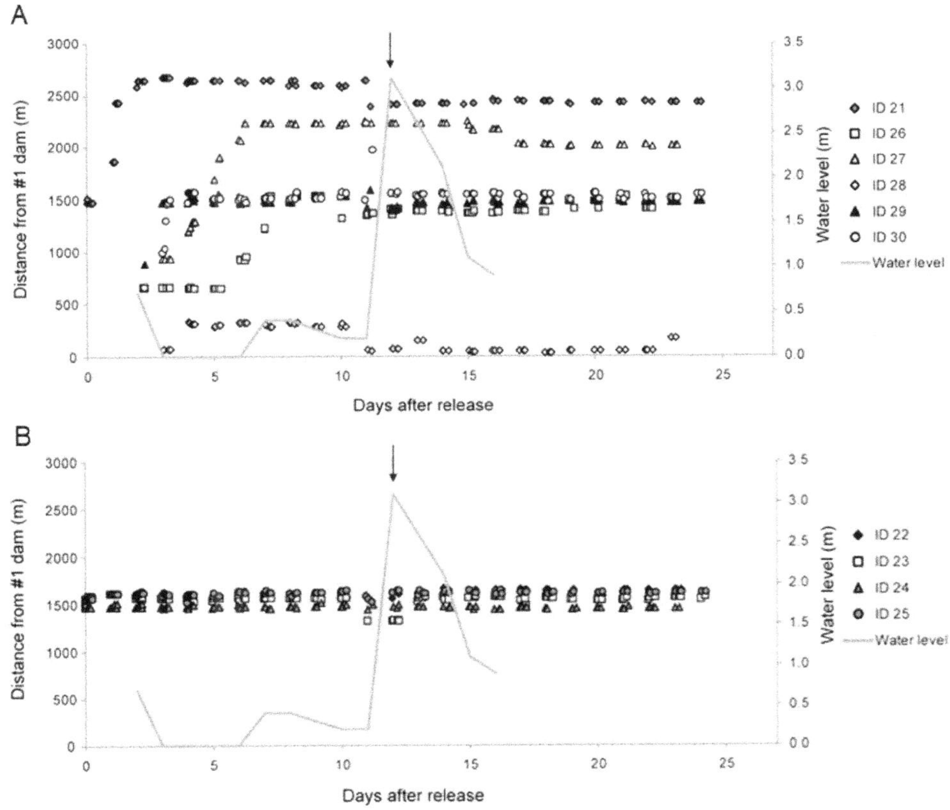

examined in a Taiwanese stream using nanotag radio telemetry in summer, particularly in the typhoon season of 2007 (Makiguchi et al., 2009b). A great flood caused by a typhoon occurred during the tracking periods in this study, but the flood had no influence on either the movements or habitat use of adult salmon. Ten tagged fish predominately used microhabitats near boulders and exhibited frequent short movements of less than 200 m between habitats (Figure 10). Our results suggest that adult salmon were basically sedentary in summer but exhibited small-scale movements to secure food. This research provides useful information for the conservation of Formosan landlocked salmon by showing that boulders in streams may serve as refuges from typhoon floods.

Comparison of the Swimming Ability and Upstream-Migration Behavior Between Chum Salmon and Masu Salmon

The spawning ground of chum salmon is usually located farther downriver than that of masu salmon in Hokkaido, Japan. To compare the swimming

Figure 11. Average (±SE) standard metabolic rate (SMR), active metabolic rate (AMR), aerobic metabolic scope (AMS), and critical swimming speed (Ucrit) of chum salmon (white column) and masu salmon (Black column). The left Y axis corresponds to oxygen consumption (MO_2 in $mgO_2h^{-1}kg^{-1}$) related to SMR, AMR, and AMS. The right Y axis corresponds to swimming speed (BLs-1) for Ucrit. Significant differences are indicated at $**p<0.01$ level.

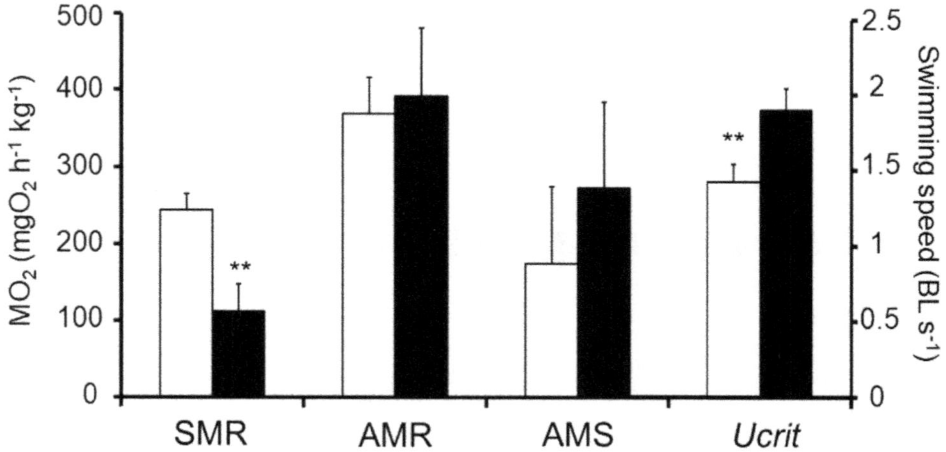

abilities of these two species, the relationship between swimming speed and oxygen consumption was compared using a swim tunnel in the laboratory. Then, the upstream migration behaviors of chum salmon and masu salmon were compared using electromyogram telemetry at fish passages in the Toyohira River, Hokkaido (Miyoshi et al., 2014). In the laboratory study, the standard metabolic rate of masu salmon was lower and the critical swimming speed (*U*crit) was faster than those of chum salmon (Figure 11). In the field study, the holding time needed to recover the swimming performance exceeding *U*crit at the fish passages and the trial number needed to pass the fish passages were significantly lower for masu salmon than chum salmon. These results revealed that masu salmon are more adaptable to extended swimming in high-water-velocity conditions than chum salmon and that masu salmon are better equipped for long-distance upstream migration to their spawning ground than chum salmon.

CONCLUSION

Due to the recent rapid development of biotelemetry technologies, the underwater behavior of salmon can be continuously observed in open water. The homing behavior of Japanese chum salmon from the central Bering Sea to Hokkaido was analyzed by using propeller data loggers that can record swimming speed, depth, and temperature over a period of 67 days over a linear distance of 2,750 km, indicating that they exhibit homeward-bound navigation abilities with a portion of time allocated to foraging. A recent magnetic archival tag study using magnetic archival tags that can record magnetic force reported that the homing migration route of Japanese chum salmon from the Bering Sea to Japan approximately followed the isoline of magnetic intensity. The homing migrations of mature lacustrine sockeye and masu salmon were tracked from the center of Lake Toya to their natal area using an ultrasonic tracking system with sensory cues impaired ablation treatment, revealed the straight homing movements of sockeye salmon using visual cues and the coastal homing movements of masu salmon using visual and olfactory cues and suggested evolutionary aspects of successful homing migration in Pacific salmon, where the narrowly distributed masu salmon required only coastal recognition ability for successful migration using olfactory and visual cues, while the widely distributed sockeye salmon required the development of open water orientation ability to successfully return to their natal streams with high precision using visual and magnetic cues. The spawning migration of lacustrine masu salmon was monitored using a swimming depth and temperature data logger in Lake Toya, which indicated that masu salmon have an innate biological clock to calculate the day length for the timing of upstream migration to their spawning site. To accurately home in open water, salmon must recognize exact locations (map) and compass directions (orientation) during migration, and they must have an internal biological clock to calculate the duration of migration from the feeding ground to the spawning ground. New research approaches that reveal how salmon can navigate in open water using compasses, maps, and biological clocks will be necessary to identify which sensory systems are involved (and how) in the oceanic imprinting and homing migration of Pacific salmon.

REFERENCES

Azumaya, T., Sato, S., Urawa, H., & Nagasawa, T. (2016). Potential role of the magnetic field on homing in chum salmon (Oncorhynchus keta) tracked from the open sea to coastal Japan. *North Pacific Anadromous Fish Commission Bulletin*, *6*(1), 235–241. doi:10.23849/npafcb6/235.241

Bracis, C., & Anderson, J. J. (2012). An investigation of the geomagnetic imprinting hypothesis for salmon. *Fisheries Oceanography*, *21*, 2–3. doi:10.1111/j.1365-2419.2012.00617.x

Cooke, S. J., Hinch, S. G., Wikelski, M., Andrews, R. D., Kuchel, L. J., Wolcott, T. G., & Butler, P. J. (2004). Biotelemetry: A mechanistic approach to ecology. *Trends in Ecology & Evolution*, *19*(6), 334–343. doi:10.1016/j.tree.2004.04.003

Cuadras, J. (1980). Cardiac responses to visual detection of movement, mechanostimulation and cheliped imposed movement in hermit crabs. *Comparative Biochemistry and Physiology. A. Comparative Physiology*, *66*(1), 113–117. doi:10.1016/0300-9629(80)90366-7

Døving, K. B., Westerberg, H., & Johnsen, P. B. (1985). Role of olfaction in the behavioral and neuronal responses of Atlantic salmon, Salmo salar, to hydrographic stratification. *Canadian Journal of Fisheries and Aquatic Sciences*, *42*(10), 1658–1667. doi:10.1139/f85-207

Formicki, K., Korezelecka-Orkisz, A., & Tanski, A. (2019). Magnetoreception in fish. *Journal of Fish Biology*, *95*, 73–91. doi:10.1111/jfb.13998

Hasegawa, E. I. (2012). Chum salmon Oncorhynchus keta respond to moonlight during homeward migrations. *Journal of Fish Biology*, *81*(2), 632–641. doi:10.1111/j.1095-8649.2012.03356.x

Hussey, N. E., Kessel, S. T., Aarestrup, K., Cook, S. J., Cowley, P. D., Fisk, A. T., ... Whoriskey, F. G. (2015). Aquatic animal telemetry: A panoramic window into the underwater world. *Science*, *348*(6240), 1255642. doi:10.1126cience.1255642

Ide, L. M., & Hoffmann, A. (2002). Stressful and behavioral conditions that affect reversible cardiac arrest in the Nile tilapia, Oreochromis niloticus (Teleostei). *Physiology & Behavior*, *75*(1-2), 119–126. doi:10.1016/S0031-9384(01)00633-3

Kaeriyama, K., & Ueda, H. (1998). Life history strategy and migration pattern of juvenile sockeye (Oncorhynchus nerka) and chum salmon (O. keta) in Japan: A review. North Pacific Anadromous Fish Commission. *Bulletin, 1*, 163–171.

King, A. J., & Adamo, S. A. (2006). The ventilatory, cardiac and behavioural responses of resting cuttlefish (Sepia officinalis L.) to sudden visual stimuli. *The Journal of Experimental Biology, 209*(6), 1101–1111. doi:10.1242/jeb.02116

Laming, P. R., & Austin, M. (1981). Cardiac responses of the anurans, Bufo bufo and Rana pipiens, during behavioral arousal and fright. *Comparative Biochemistry and Physiology. A. Comparative Physiology, 68*(3), 515–518. doi:10.1016/0300-9629(81)90084-0

Lin, Y. S., Tsao, S. S., & Chang, K. H. (1990). Population and distribution of the Formosan landlocked salmon (Oncorhynchus masou formosanus) in Chichiawan Stream. *Bulletin of the Institute of Zoology, Academia Sinica, 29*(Supplement 3), 73–85.

Makiguchi, Y., Liao, L. Y., Konno, Y., Nii, H., Nakao, K., Gwo, J. C., ... Ueda, H. (2009b). Site fidelity of and habitat use by the Formosan landlocked salmon (Oncorhynchus masou formosanus) during typhoon season in Chichiawan stream, Taiwan as assessed by nano-tag radio telemetry. *Zoological Studies (Taipei, Taiwan), 48*, 460–467.

Makiguchi, Y., Nagata, S., Kojima, T., Ichimura, M., Konno, Y., Murata, H., & Ueda, H. (2009a). Cardiac arrest during gamete release in chum salmon regulated by the parasympathetic nerve system. *PLoS One, 4*(6), e5993. doi:10.1371/journal.pone.0005993

Makiguchi, Y., Nii, H., Nakao, K., & Ueda, H. (2008). Migratory behaviour of adult chum salmon, Oncorhynchus keta, in a reconstructed reach of the Shibetsu River, Japan. *Fisheries Management and Ecology, 15*(5-6), 425–433. doi:10.1111/j.1365-2400.2008.00632.x

Miyoshi, K., Hayashida, K., Sakashita, T., Fujii, M., Nii, H., Nakao, K., & Ueda, H. (2014). Comparison of the swimming ability and upstream-migration behavior between chum salmon and masu salmon. *Canadian Journal of Fisheries and Aquatic Sciences, 71*(2), 217–225. doi:10.1139/cjfas-2013-0480

Ogura, M., & Ishida, Y. (1994). Homing behavior and vertical movements of four species of Pacific salmon (Oncorhynchus spp.) in the central Bering Sea. *Canadian Journal of Fisheries and Aquatic Sciences*, *52*(3), 532–540. doi:10.1139/f95-054

Putman, N. F., Jenkins, E. S., Michielsens, C. G. J., & Noakes, D. L. G. (2013). Geomagnetic imprinting predicts spatio-temporal variation in homing of pink and sockeye salmon. *Journal of the Royal Society, Interface*, *11*(99), 20140542. doi:10.1098/rsif.2014.0542

Quinn, T. P., & Groot, C. (1984). Pacific salmon (Oncorhynchus) migrations: Orientation vs. random movement. *Canadian Journal of Fisheries and Aquatic Sciences*, *41*(9), 1319–1324. doi:10.1139/f84-161

Quinn, T. P., Terjart, B. A., & Groot, C. (1989). Migratory orientation and vertical movements of homing adult sockeye salmon, Oncorhynchus nerka, in coastal waters. *Animal Behaviour*, *37*, 587–599. doi:10.1016/0003-3472(89)90038-9

Tanaka, H., Naito, Y., Davis, N. D., Urawa, S., Ueda, H., & Fukuwaka, M. (2005). First record of the at-sea swimming speed of a Pacific salmon during its oceanic migration. *Marine Ecology Progress Series*, *291*, 307–312. doi:10.3354/meps291307

Ueda, H. (2004). Recent biotelemetry research on lacustrine salmon homing migration. Memoirs of National Institute of Polar Research, 58(SI), 80–88.

Ueda, H. (2014). Homing ability and migration success in Pacific salmon: Mechanistic insights from biotelemetry, endocrinology, and neurophysiology. *Marine Ecology Progress Series*, *496*, 219–232. doi:10.3354/meps10636

Ueda, H. (2016). Physiological mechanisms of imprinting and homing migration of Pacific salmon. *Aqua-BioScience Monographs*, *9*(1), 1–27. doi:10.5047/absm.2016.00901.0001

Ueda, H., Kaeriyama, M., Mukasa, K., Urano, A., Kudo, H., Shoji, T., ... Kurihara, K. (1998). Lacustrine sockeye salmon return straight to their natal area from open water using both visual and olfactory cues. *Chemical Senses*, *23*(2), 207–212. doi:10.1093/chemse/23.2.207

Urawa, H. (2004). Stock identification studies of high seas salmon in Japan: A review and future plan. *North Pacific Anadromous Fish Technical Report*, *5*, 9–10.

Walker, M. M., Diebel, C. E., Haugh, C. V., Pankhurst, P. M., Montgomery, J. C., & Green, C. R. (1997). Structure and function of the vertebrate magnetic sense. *Nature*, *390*(6658), 371–376. doi:10.1038/37057

Yano, A., Ogura, M., Sato, A., Sakaki, Y., Ban, M., & Nagasawa, K. (1996). Development of ultrasonic telemetry technique for investigating the magnetic sense of salmonids. *Fisheries Science*, *62*(5), 698–704. doi:10.2331/fishsci.62.698

Yano, K., & Nakamura, A. (1992). Observations of the effect of visual and olfactory ablation on the swimming behavior of migrating adult chum salmon, Oncorhynchus keta. *Japanese Journal of Ichthyology*, *39*(1), 67–83. doi:10.1007/BF02905635

Chapter 6
Salmon Resources

ABSTRACT

Salmon are among the most important fishery resources and are produced by commercial fisheries, aquaculture, and propagation. The total production of salmon is now 4.6 million tons, among which 22% comes from commercial fishery production and 78% from aquaculture production, which is the fastest growing food production system in the world. Atlantic salmon and rainbow trout are produced by marine and freshwater aquaculture mainly in Norway and Chile, where environmental impacts are intensively investigated. The biomass of Japanese chum salmon increased steadily from 1970 to 1996 because of the successful improvement of propagation systems. However, the return rate of homing adults has decreased sharply, mainly due to recent unpredictable climate change. New trials of Japanese salmon propagation systems are being carried out by using semi-closed recirculating aquaculture systems and orally administering docosahexaenoic acid.

INTRODUCTION

Salmon resources are produced by commercial fisheries, marine and freshwater aquaculture (farming), and propagation (sea ranching: juvenile release from the hatchery in their natal stream and adult catches in their natal area several years later). The salmon commercial fishery used to consist of free catches from fishing vessels using drift net or longline fishing in the North Pacific and Atlantic Oceans. An exclusive economic zone (EEZ) extending 200 nautical miles from the coast was prescribed by the 1982

DOI: 10.4018/978-1-7998-2054-3.ch006

United Nations Conservation on the Law of Sea for the special right to explore and use marine resources. Since then, an intergovernmental organization, the North Atlantic Salmon Conservation Organization (NASCO), was established in 1983 to create a large protected zone free of targeted fisheries for Atlantic salmon in most areas beyond 12 nautical miles from the coast by Canada, Denmark (with respect to the Faroe Islands & Greenland), the European Union, Norway, the Russian Federation, and the United States of America. The North Pacific Anadromous Fish Commission (NPAFC) was established in 1993 to promote the conservation of anadromous stocks of Pacific salmon in the convention area including the international waters of the North Pacific Ocean and its adjacent seas north of 33° North beyond the 200-mile EEZ of the coastal seas of Canada, Japan, the Republic of Korea, the Russian Federation, and the United States of America.

Almost all commercial fishery production of salmon consists of Pacific salmon because the capture of Atlantic salmon has only been a few thousand tons in recent years. The catch of Pacific salmon was approximately 0.4 million tons with small yearly fluctuations until the 1970s, and it then increased up to one million tons with large yearly fluctuations in recent years, mainly in the United States of America (USA), the Russian Federation (USSR until 1987), Japan, and Canada (Figure 1A). The largest commercial salmon fishery production is observed for pink salmon, with large differences in odd and even years; the second largest is for chum salmon; the third largest is for sockeye salmon; and the fourth largest is for coho salmon (Figure 1B). Commercial salmon fishery production includes salmon propagation systems in each country.

Total salmon aquaculture production has been 3.6 million tons in recent years. Salmon marine aquaculture (mariculture) production has reached 3 million tons, mainly in Norway and Chile, with some production also coming from Iran, the United Kingdom, Canada, the Faroe Islands, Australia, and Denmark (Figure 2A). The largest marine aquaculture species is Atlantic salmon, the second largest is steelhead (so-called Salmon trout), the third largest is coho salmon, and the fourth largest is Chinook salmon (Figure 2B). Recently, salmon freshwater aquaculture production has reached approximately 0.6 million tons in Iran, Turkey, France, Italy, Denmark, the USA, Germany, Spain, Japan, Chile, and China (Figure 3A). The main freshwater aquacultured species are rainbow trout and, to a lesser extent, Atlantic salmon (Figure 3B).

Figure 1. Annual changes in the Pacific salmon fishery production of different countries (A) and species (B) from 1950 to 2016. Data source: FishStatJ of FAO (2018a), figure produced by Dr. Ikutaro Shimizu.

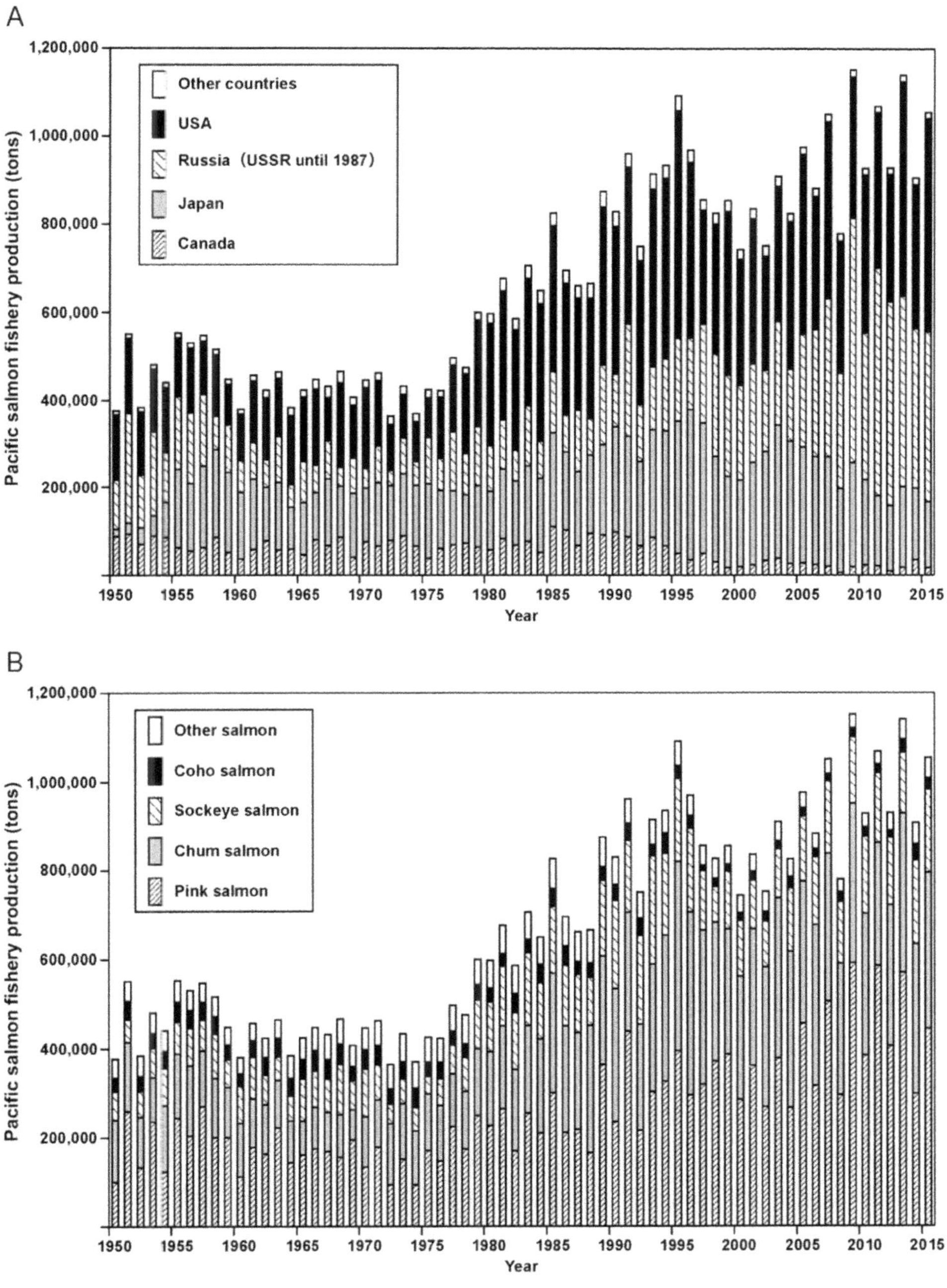

The total production of salmon is now 4.6 million tons, 22% of which comes from commercial fishery production and 78% from aquaculture production. Salmon aquaculture is the fastest growing food production system in the world, showing over 30-fold growth over 30 years due to improvements in formulated feeds, the biological understanding of farmed salmon, and the demand for seafood products. However, salmon aquaculture has been subject to many concerns regarding environmental and social impacts, such as water pollution, genetic interaction of nonnative species with wild fish, escapes, the spread of diseases, and unfair labor practices at salmon farms.

Salmon Aquaculture

Salmon aquaculture had begun as freshwater aquaculture using rainbow trout in the United States of America in the late 19th century and has grown dramatically since the 1950s. In recent years, steelhead have been produced in mariculture in Chile and Norway (Figure 2A, B) and rainbow trout have been produced in freshwater aquaculture in many countries (Figure 3A, B). Atlantic salmon mariculture started in Norway in the 1980s and in Chile in the 1990s and has recently begun in Canada the United Kingdom, Ireland, the Faroe Islands, the Russian Federation, and Tasmania. Coho salmon mariculture is carried out in Chile, Canada, and Japan. Chinook salmon mariculture and freshwater aquaculture is conducted in New Zealand.

Salmon mariculture takes place first in freshwater hatcheries, and the fish are then transferred to sea cages (sea pens or net pens). Freshwater, inland, hatcheries produce and supply salmon smolts that exhibit seawater adaptability. In the past, hatcheries operated flow-through systems using natural spring water or river water and have recently shifted to the use of recirculating aquaculture systems (RASs) in which the water is recycled, and water temperature and water flow are controlled to modulate the growth rate of smolts to meet the needs of sea cage systems. In hatcheries, artificial fertilized eggs are produced from mature males and females by artificial insemination, and hatching fry are reared with dry pellet feed in freshwater ponds until the smolts are transported to sea cages.

Sea cages are large floating net cages of approximately 10~30 m in diameter and 10~15 m in depth with a circular, octagonal, or square shape and a volume of 1,000~10,000 m^3 that are moored to buoys, usually in calm, clean seawater along the coastal areas of gulfs, such as fjord and rias

Figure 2. Annual changes in the salmon marine aquaculture production of different countries (A) and species (B) from 1950 to 2016. Data source: FishStatJ of FAO (2018b), figure produced by Dr. Ikutaro Shimizu.

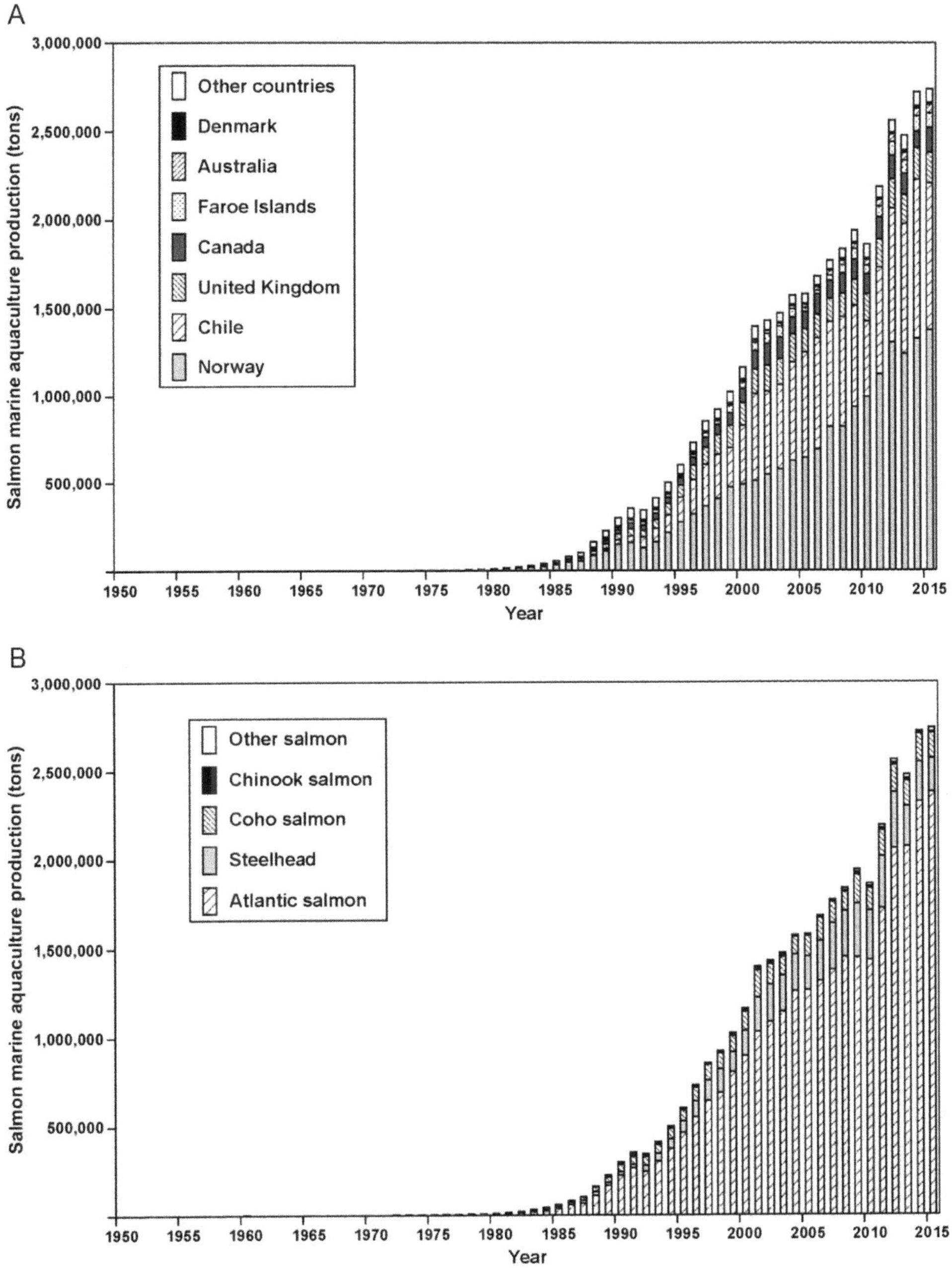

Figure 3. Annual changes in salmon freshwater aquaculture production of different countries (A) and species (B) from 1950 to 2016. Data source: FishStatJ of FAO (2018b), figure produced by Dr. Ikutaro Shimizu.

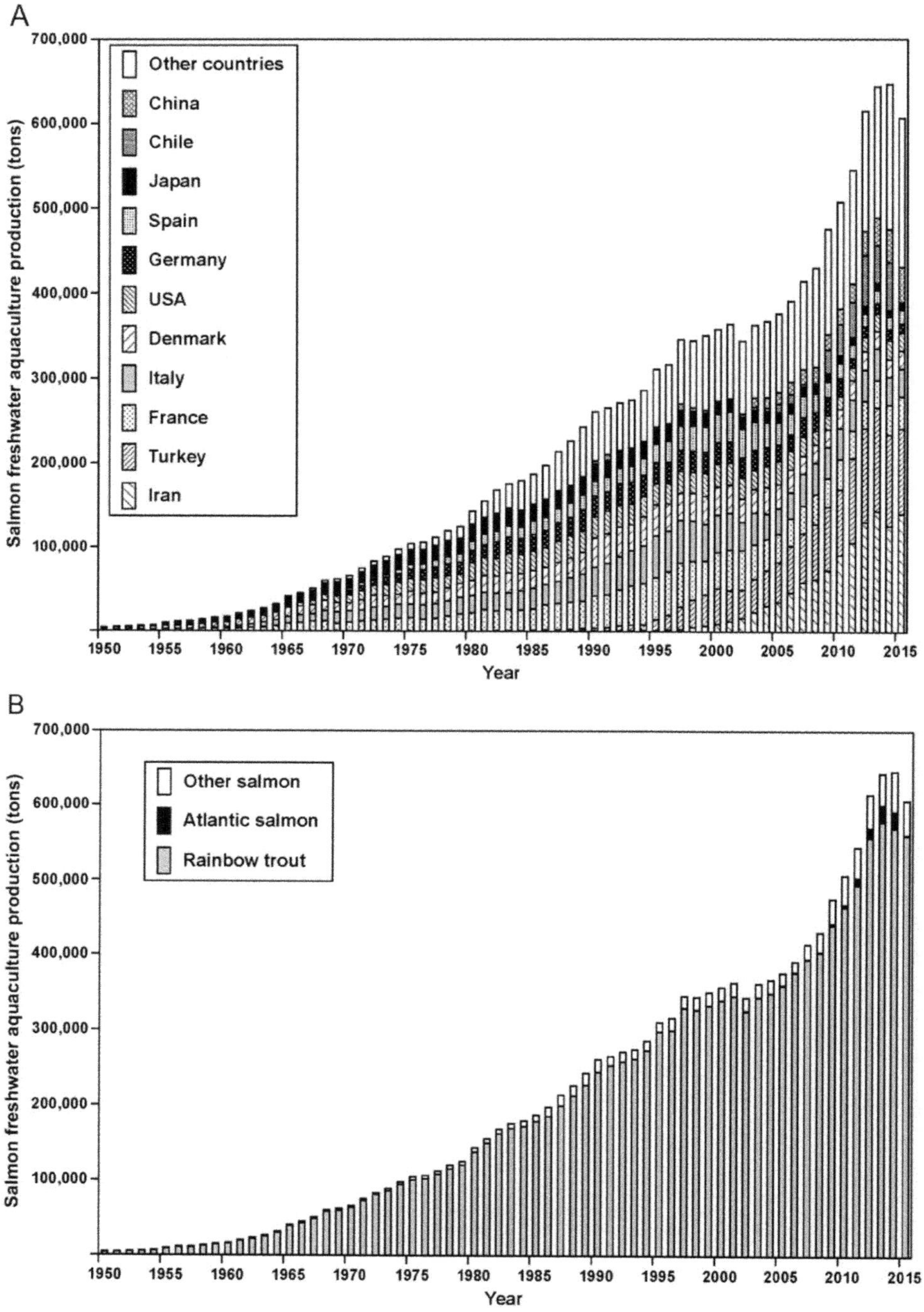

coasts. Copper alloys can be used as antimicrobial netting materials to deter bacteria, viruses, fungi, algae, and other microbes. In sea cages, salmon are fed fish feeds consisting mainly of fishmeal, fish oil, and soybean meal. To produce 1 kg of farmed salmon, 2~4 kg of wild-caught fish is needed to prepare the fish feed (Naylor et al., 1998). With the decline in global fishery catches since the late 1980s, the continuation of present trends will lead to supply shortfalls for which aquaculture cannot be expected to compensate, and the situation may well be exacerbated (Pauly & Zeller, 2005).

Environmental Impacts of Aquaculture in Norway and Chile

Norway and Chile are two major salmon farming nations in which environmental impacts have been intensively investigated. Norway is a pioneering country in salmon farming that has been considered to have the potential to influence the surrounding environment and wild populations. The risk assessment of the environmental impact of Norwegian Atlantic salmon farming has been repeatedly performed by assessing the genetic introgression of farmed salmon in wild populations, the regulatory effects of salmon lice and viral diseases on wild salmonid populations, and the local and regional impact of nutrients and organic loading (Taranger et al., 2015). These studies have indicated (i) a moderate-to-high risk of genetic introgression from farmed escaped salmon in some wild salmon populations, (ii) a moderate-to-high risk of mortality from salmon lice infection in salmon smolts and wild sea trout, (iii) the occurrence of viral disease outbreaks caused by the extensive release of viruses in many areas, and (iv) a low risk of eutrophication and organic loading under salmon farming.

In contrast, the eutrophication of Patagonian channels and fjords caused by nutrients derived from marine salmon farms has been recognized as a crucial environmental risk triggering harmful algal blooms in Chile. In addition, the impacts of local eutrophication and changes in diversity downstream of freshwater hatcheries that are deleterious to some nontarget species caused by pesticides used to control sea louse infestations and effects of increases in antibiotic-resistant bacteria in the sediments of salmon farms and the proportion of antimicrobial-resistant bacteria have also been recognized. There is an urgent need for more comprehensive ecosystem studies on the impact of antibiotics (Quinones et al., 2019).

In Norway, genetic investigations have focused on progress resulting from genome-wide studies aimed at improving aquaculture production and the understanding of salmonid ecology, physiology, and evolution by means of emerging technologies such as genome editing (Houston & Macqueen, 2019). In Chile, genetic improvement efforts have focused on the development of efficient aquaculture based on genetic parameters and responses to selection and the incorporation of genomics (Lhorente et al., 2019). These genetic studies should focus on the food safety of genetically modified farmed salmon.

Marine Ecolabeling

The Food and Agriculture Organization of the United Nations (FAO) established Guidelines for the Ecolabeling of Fish and Fishery Products from Marine Capture Fisheries in 2009, which were designed to certify and promote the labeling of products from well-managed marine capture fisheries with a focus on issues related to the sustainable use of fisheries resources (FAO, 2009). Following these guidelines, the World Wide Fund for Nature (WWF) produced an environmental and social standard for farmed salmon and funded the Aquaculture Stewardship Council (ASC), which is an independent nonprofit organization, to establish a protocol for farmed seafood while ensuring sustainable aquaculture in 2010. In 2017, the ASC identified and characterized seven key environmental and social principles of salmon aquaculture (Table 1) (ASC, 2017). The mission of the ASC is to transform aquaculture towards environmental sustainability and social responsibility using efficient market mechanisms that create value across the chain. The Marine Stewardship Council (MSC) was also founded by the WWF in 1996. The mission of the MSC is to use ecolabeling and fishery certification programs to contribute to the health of the world's oceans by recognizing and rewarding sustainable fishing practices, influencing the choices people make when buying seafood and working with our partners to transform the seafood market on a sustainable basis. Recently, the Marine Eco-label Japan Council (MEL) has established the aim of increasing social demand for nations, corporations and individuals to demonstrate a high standard of ethical behavior through the mission of the fisheries industry to ensure the sustainable use of fishery resources and conserve the marine ecosystem.

Table 1. Seven key environmental and social principles of salmon aquaculture

	Principle		Criterion
1	Comply with all applicable national laws and local regulations	1.1	Compiance with all applicalbe local and national legal requirements and regulations
2	Conserve natural habitat, local biodiversity and ecosystem function	2.1	Benthic biodiversity and benthic effects
		2.2	Water quality in and near the site of operation
		2.3	Nutrient release from production
		2.4	Interaction with critical or sensitive habitats and species
		2.5	Interaction with wildlife, including predators
3	Protec the health and genetic integrity of wild populations	3.1	Introduced or amplified parasites and pathogens
		3.2	Introduction of non-native species
		3.3	Introduction of transgenic species
		3.4	Escapes
4	Use resources in an environmentally efficient and responsible manner	4.1	Traceability of raw materials in feed
		4.2	Use of wild fish for feed
		4.3	Source of marine raw materials
		4.4	Source of non-marine raw materials in feed
		4.5	Non-biological waste from production
		4.6	Energy consumption and greenhouse gas emissions on farms
		4.7	Non-therapeutic chemical inputs
5	Manage disease and parasites in an environmentally responsible manner	5.1	Survival and health of farmed fish
		5.2	Therapeutic treatments
		5.3	Resistance of parasites, viruses and bacteria to medicinal treatments
		5.4	Biosecurity management
6	Develop and operate farms in a socially responsible manner	6.1	Freedom of association and collective bargaining
		6.2	Child labor
		6.3	Forced, bonded or compulsory labor
		6.4	Discrimination
		6.5	Work environment health and safety
		6.6	Wages
		6.7	Contracts (labor) including subcontracting
		6.8	Conflict resolution
		6.9	Disciplinary practices
		6.1	Working hours and overtime
		6.11	Education and training
		6.12	Corporate policies for social responsibility
7	Be a good neighbor and conscientious citizen	7.1	Community engagement
		7.2	Respect for indigenous and aboriginal cultures and traditional territories
		7.3	Access to resources

SALMON PROPAGATION SYSTEMS IN JAPAN

History and Transition

Artificial salmon/trout propagation techniques involving juvenile release from hatcheries into the natal stream and adult catches in the natal area several years later were introduced from Western countries in the late 1870s. Akikiyo Sekizawa was impressed by artificial salmon/trout propagation techniques observed both in Weltausstellung, Wien, Austria, in 1873 and at the International Exhibition of 1876 in Philadelphia, USA, and carried out the first trial juvenile chum salmon release in the Nakagawa River, Ibaraki Prefecture in 1876. Full-scale chum salmon propagation was initiated in 1888, when the first governmental hatchery was implemented in the Chitose

River, Hokkaido, mainly due to the efforts of Kazutaka Ito, who studied an artificial salmon propagation system at the Bucksport hatchery in Maine, USA, from 1886-1887.

Due to the implementation of the Fisheries Resource Protection Law in 1951 in Japan and the enormous efforts made to establish juvenile release techniques in the 1960s, such as employing Atkins incubators for large numbers of artificially fertilized eggs, the number of returning adult chum salmon reached 3-5 million by 1965. Coastal surveys of juveniles were first conducted in 1969 and showed that the timing of juvenile release to when the coastal sea surface temperature (SST) was approximately 5-10°C led to the highest survival rate of juveniles. This favorable timing of juvenile release as well as improved rearing techniques for producing many healthy juveniles by feeding dry pellets to fry following yolk absorption in the late 1970s caused steady increases in the numbers of juveniles released from 0.5 to 2 billion and returning adults from 10 to 90 million from the late 1970s to 1996 (Figure 4).

However, despite the almost constant release of approximately 1.8-2 billion juveniles, a sharp decline in adult returns was observed with an approximately 8-year interval from 1992 to 2008. The mechanisms underlying the 8-year interval between juvenile release and adult return have not yet been clarified; however, salmon biomass is negatively affected by long-term climate change in the North Pacific Ocean, which has experienced climate change with decadal oscillations (Kaeriyama et al., 2014). Global warming has also positively affected the survival rate of juvenile chum salmon in the Okhotsk Sea and caused population density-dependent effects on growth from ages 2 to 4 and maturation in the Bering Sea (Seo et al., 2011). Kitada (2014) has reviewed the present status of the Japanese chum salmon stock enhancement program and suggested the importance of juvenile quality and the vital roles of escapements in enhanced and nonenhanced rivers.

In 2003, an amendment to the Pharmaceutical Affairs Law prohibited the use of formaldehyde, which had been quite effective in protection against fish diseases such as bacterial gill disease and ichthyobodosis. No other effective methods for protection against various fish diseases have been developed to date. Moreover, a major decrease in juvenile release from hatcheries was caused by the aftermath of the Tohoku earthquake and tsunami in 2011. After this major damage, although the number of hatchery juveniles released recovered to 1.75 billion, the commercial adult catch decreased sharply to 20 million fish (Figure 4). Although the optimal juvenile release criteria

Figure 4. Annual changes in the numbers of hatchery juveniles released (line) and commercial adult catches (pale bar in Northern Honshu and dark bar in Hokkaido) in Japan from 1970 to 2017). The data were provided by the Hokkaido National Fisheries Research Institute.

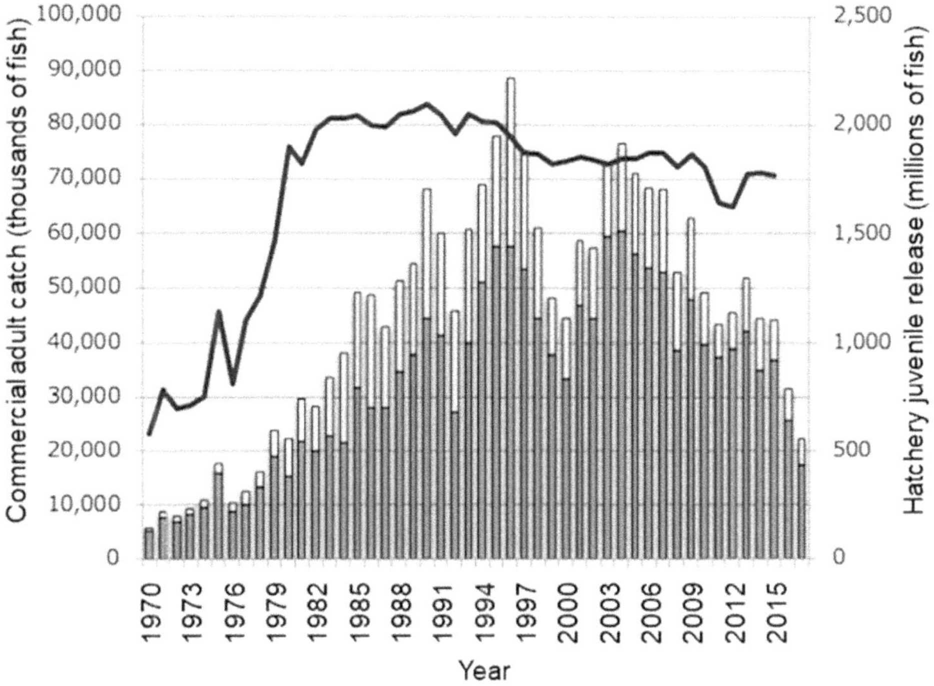

of an approximately 1 g juvenile body weight and coastal SSTs between 5 and 13°C were established in the late 1990s, recent unpredictable climate change has had profound effects on the survival of Japanese chum salmon populations in the North Pacific Ocean (Ueda, 2015).

Current Propagation Systems

Artificially fertilized eggs are stocked for one month in box-type incubators that can hold half a million eggs, which are supplied with clean cool spring water. Dead eggs are removed with a picking machine after the eyed stage, and live eggs are disinfected with iodine solution to prevent viral and bacterial diseases. Eyed eggs are incubated with cooled spring water at approximately 4°C to induce thermal otolith marking at the hatchery of the Hokkaido National Fisheries Research Institute. Several days before hatching, eyed eggs are placed in hatchery trays in gravel raceways, where the hatchery trays are set to maintain convectional water exchange. The hatchery trays are removed from the gravel raceways after hatching. The hatched alevins are left in the gravel raceways for approximately 2 months at 8°C to absorb nutrients in their yolk sac. Chum salmon fry are transferred from the gravel raceways to outdoor feeding ponds following yolk absorption, approximately 2 months after hatching, and are fed with dry pellets for 1-2 months. At a body weight of approximately 1-2 g, juveniles are released into rivers between late March and mid-April. The juveniles then migrate to the sea after a short stay in the river (Figure 5).

The migration route of the Japanese chum salmon was estimated using information on fish abundance and data from the genetic identification of stocks sampled in the open ocean from 1993 to 1999 (Figure 1 in Chapter 5) (Urawa, 2004). Juvenile chum salmon of less than 1 year of age are distributed in the Okhotsk Sea from August to November but are confined to a narrow band (SST 4–6°C) of the western North Pacific from December to May. One-year-old salmon enter the Bering Sea from June to November. In the late fall, they travel southeast to the Gulf of Alaska, where they remain from December to May. These fish migrate between their summer feeding grounds in the Bering Sea and their winter habitat in the Gulf of Alaska for 3-4 years until they begin their homing migration for spawning, triggered by the onset of gonadal maturation in the Bering Sea and western North Pacific.

Adult salmon are caught using set nets in coastal fisheries along the coastline of Hokkaido and northern Honshu in both the Japanese Sea and Pacific Ocean. Escaped adults continue their upstream migration and are caught in weirs, usually at the mouths of rivers. The fish are then transferred to the hatchery, reared until final gonadal maturation occurs, and used for artificial insemination. New chum salmon propagation techniques await development to improve both the survival rate of juveniles and the return rate of adult chum salmon. It is important to clarify the mechanisms of imprinting and homing migration in chum salmon to improve the survival rates of juveniles and return rates of adults.

Figure 5. The current chum salmon propagation system used in Japan, showing the cycle of the salmon enhancement program operated by the Hokkaido Salmon Propagation Association.

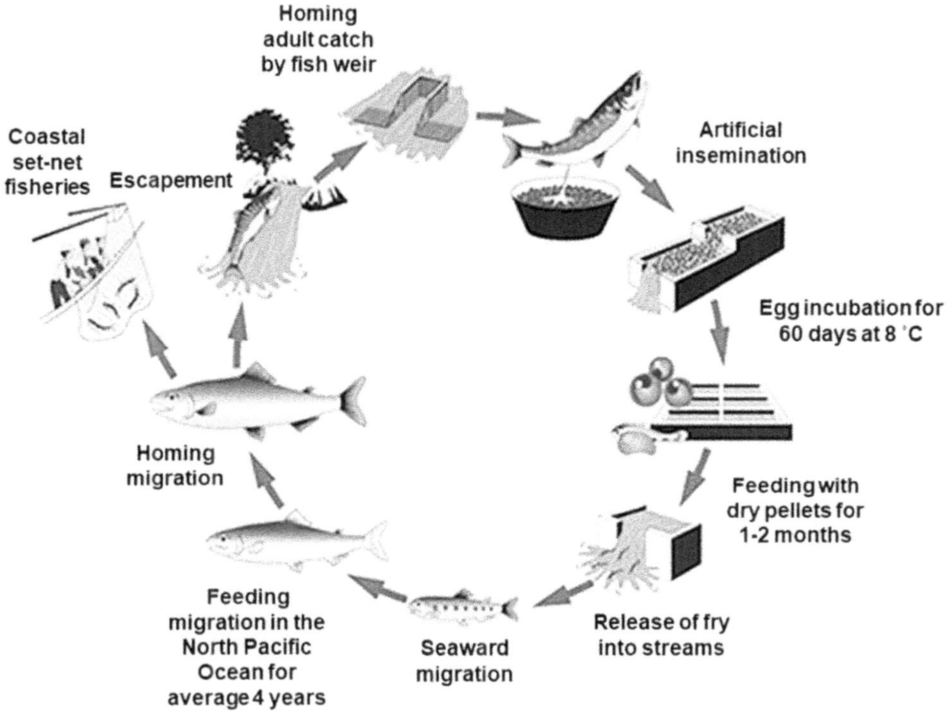

New Trials of Japanese Salmon Propagation Systems

To achieve long-term sustainable chum salmon production in Japan, high rates of both juvenile survival and adult return are crucial. To produce healthy juveniles with high survival rates, a new chum salmon propagation system should be developed by introducing semiclosed recirculating aquaculture systems, estimating the health condition of fish, and improving the olfactory imprinting ability of juvenile chum salmon.

Semiclosed Recirculating Aquaculture System

In most chum salmon hatcheries in Japan, chum salmon fry are reared by feeding them commercial pellets in square concrete ponds until they are

Figure 6. A semi-closed aquaculture system for juvenile chum salmon consisting of a bio-active filtration tank, an air blower in the filtration tank, an ultraviolet sterilization instrument, two water pumps, two Coolnics circulators, and two autofeeders, all within an oval-shaped fiber-reinforced plastic fish tank.

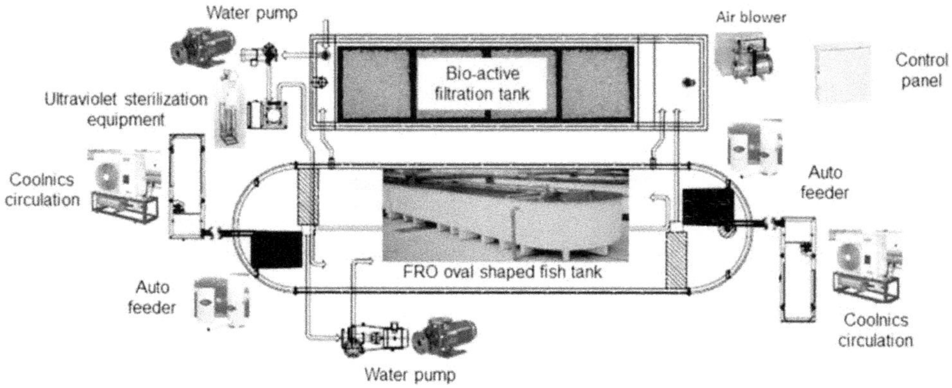

released into the stream. These ponds are supplied with spring water at a constant flow rate of 0.8-1.1 cm/sec, usually creating dead water space at each corner and are often prone to fish diseases such as bacterial gill disease and ichthyobodosis. Moreover, the shortage of good-quality spring water has become a serious problem at hatcheries in Japan. To overcome these problems, we are developing a semiclosed RAS for juvenile chum salmon (Figure 6). This RAS consists of a bioactive filtration tank to maintain high-quality water, an air blower in the filtration tank, an ultraviolet sterilization system for the disinfection of water, two water pumps to control water circulation, two Coolnics circulators to control water temperature, and two autofeeders to feed pellets containing bioactive substances, all within an oval-shaped, fiber-reinforced plastic (FRP) fish tank. Using this new system, we are now examining the culture of high-quality healthy juvenile chum salmon. A preliminary evaluation of water quality according to NH_3 and HNO_2 levels showed no differences between the RAS and standard salmon hatchery pond water. In addition, preliminary analysis of the growth rates of chum salmon revealed no differences between the RAS and the standard salmon hatchery pond, suggesting that the RAS does not subject the salmon to chronic stress.

Estimation of Health Condition

Mizuno et al. (2010a, b) examined the relationship between rearing conditions and the health of juvenile chum salmon using body adenosine triphosphate (ATP) contents and ATP synthesis-related transcript levels as health parameters and found that these parameters could be used to effectively evaluate the health conditions of juvenile chum salmon. The health conditions of these fish are strongly correlated with their metabolism and swimming ability. Improvement of the metabolism and swimming ability of juvenile chum salmon can directly contribute to increased survival rates in coastal areas following downstream migration. Using the RAS, we are investigating the effects of controlling the water circulation speed on the improvement of the metabolism and swimming ability of juvenile chum salmon.

Improvement of the Olfactory Imprinting Ability of Masu Salmon

A trial experiment was conducted to improve the olfactory imprinting ability of masu salmon smolts via the oral administration of three types of bioactive substances: docosahexaenoic acid (DHA), NMDA, and T4. DHA has been reported to improve memory learning in the mouse brain (Hiratsuka et al., 2009); NMDA is involved in the development of neural networks, learning and memory in chicks (Nakamori et al., 2013); and T4 determines the onset of the sensitive period for imprinting in chicks (Yamaguchi et al., 2012). Preliminary results showed that these three bioactive substances increased NR1 and TRHa/b gene expression (although T4 administration did not increase TRHb expression), suggesting that these bioactive substances might affect the imprinting ability of masu salmon smolt (Figure 7) (Ueda, 2015). In addition, anadromous salmonid fish showed a distinct surge in plasma T4 levels coinciding with the new moon during the smoltification period prior to downstream migration (Grau et al., 1980; Yamauchi et al., 1985).

The effects of the oral administration of pellets containing DHA on the homing rate of adult masu salmon were examined. Masu salmon smolts were tagged with PIT tags and released from the Shimoakka Hatchery on the day of the new moon, and their return to the native hatchery was monitored via the PIT tag detection system. The Iwate Prefecture Inland Fisheries Technology Center monitored the homing rate of adult masu salmon in the Akka River via the fin-clipping and ribbon-tagging of smolts released

*Figure 7. Preliminary results of the effects of the oral administration of bioactive substances (T4, DHA, and NMDA) on gene expression levels of NMDA receptor essential subunit NR1 and TRHa/b in juvenile masu salmon. Significant differences are indicated at *p<0.05 level.*

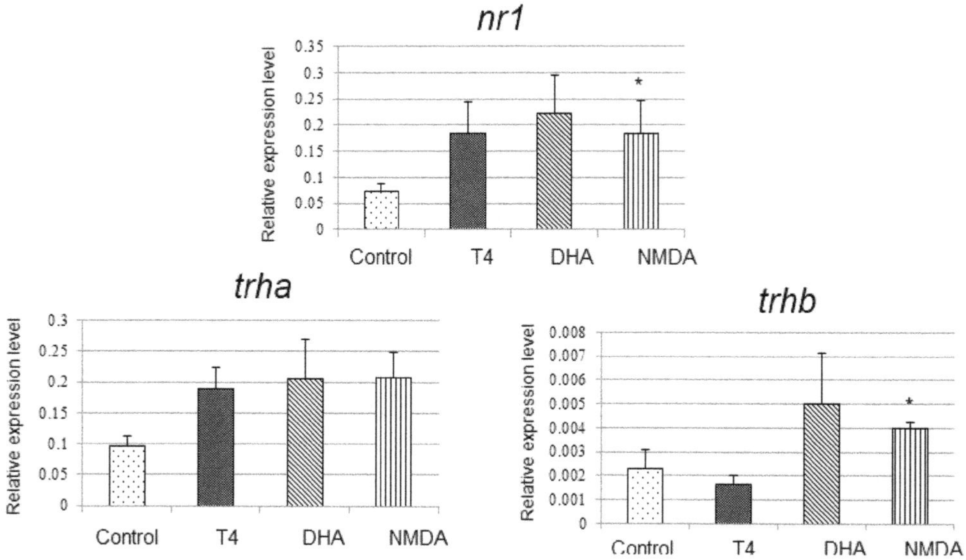

from 1996 to 2004 and reported homing rates (%) ranging from 0.001 to 0.118 (Otomo et al., 2006). The homing rates of DHA-administered adult masu salmon were 0.333% (5/1500) in 2014, 1.177% (15/1274) in 2015, and 0.511 (6/1174) in 2016, which constituted a 3~10-fold increase over the previous highest homing rate at the Shimoakka Hatchery (Figure 8).

These results demonstrate that the oral administration of DHA to smolts most likely increases the homing rate of adult masu salmon. There are three possible explanations for this increase in the adult masu salmon homing rate. The first explanation involves the enhancement of imprinting ability through an increase in *nr1* and *trha/b* gene expression levels in the masu salmon brain, which may decrease the straying rate of adult masu salmon. The second explanation involves the enhancement of the survival rate of masu salmon smolts immediately after release in the coastal sea due to improved seawater adaptability in the osmoregulatory organs. The third explanation involves the enhancement of swimming activity during feeding migration and homing migration to avoid predator attacks on muscle tissues. To

Figure 8. Homing rates of adult masu salmon in the Shimoakka Hatchery from 2014 to 2016. The PIT tag detection system was washed away by Typhoon 10 on August 30, 2016.

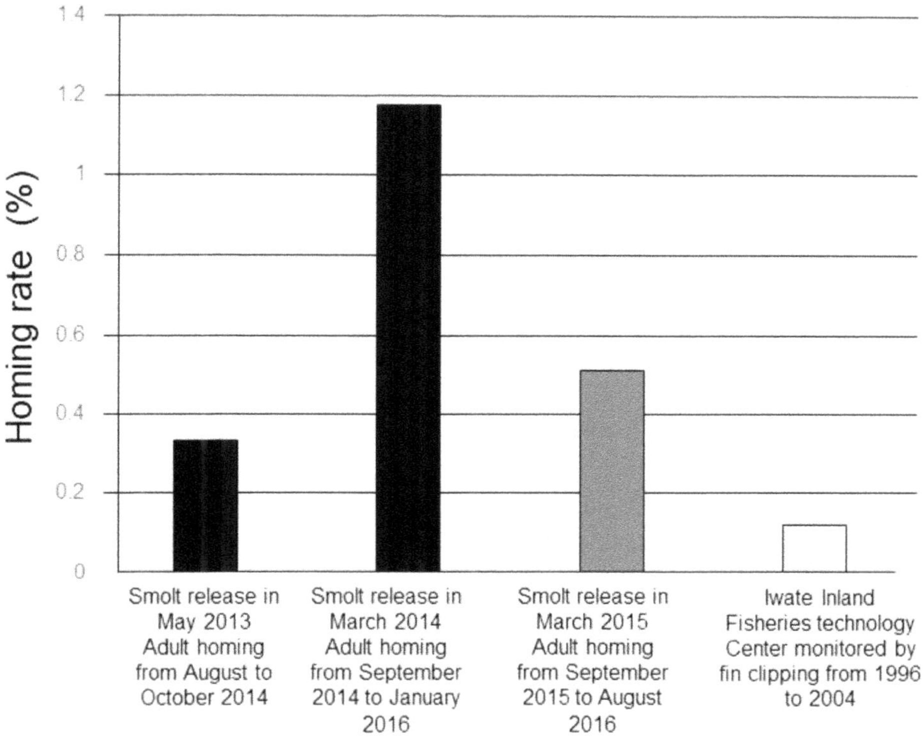

discriminate between these possibilities, further physiological investigations should be performed to investigate the mechanisms underlying the increased homing rate in adult masu salmon following oral DHA administration.

CONCLUSION

Salmon are very important fishery resources and are mainly produced by commercial fisheries, aquaculture and propagation. Since the EEZ was prescribed in 1982, NASCO and NPAFC have controlled commercial catches of Atlantic and Pacific salmon, respectively. Almost all commercial fishery production of salmon consists of Pacific salmon (1st pink, 2nd chum,

3rd sockeye, and 4th coho salmon), for which recent production has been one million tons with large yearly fluctuations due to differences in pink salmon between odd and even years. Salmon aquaculture is the fastest growing food production system in the world, showing 30-fold growth for 30 years, with 3 million tons now being produced through marine aquaculture (mariculture) and 0.6 million tons through freshwater aquaculture. Atlantic salmon, steelhead, and coho salmon are mainly produced via mariculture in Norway and Chile. Repeated risk assessments of the environmental impact of Norwegian Atlantic salmon farming the genetic introgression of farmed salmon in wild populations, the regulatory effects of salmon lice and viral diseases on wild salmonid populations, and the local and regional impact of nutrients and organic loading. In Chile, the eutrophication of Patagonian channels and fjord caused by nutrients derived from marine salmon farms has been recognized as a crucial environmental risk for the triggering of harmful algal blooms. Due to the economic and environmental importance of salmon aquaculture, the ASC established seven key environmental and social principles of salmon aquaculture. In Japan, since salmon propagation techniques were introduced from Western countries in the late 1870s, chum salmon have mainly been propagated by artificial insemination, the release of juveniles from their natal river to the ocean, and the recapture of homing adults along the coast and within the natal river. The biomass of Japanese chum salmon increased steadily from 1970 to 1996 due to enormous efforts to establish juvenile release techniques. Although the optimal juvenile release criteria of an approximately 1 g juvenile body weight and coastal SSTs between 5 and 13°C were be established in the late 1990s, recent unpredictable climate change caused a decrease in production from 90 million tons in 1996 to 20 million tons in 2017. New trials of Japanese salmon propagation systems are being implemented using the semiclosed RAS, and no differences in the growth rates of juvenile chum salmon have been observed between the RAS and standard salmon hatchery ponds. The effect of the oral administration of pellets containing DHA to masu salmon smolts on the homing rate of adult masu salmon in the Shimoakka Hatchery, Iwate Prefecture, revealed a 3~10-fold increase over the previous highest homing rate at the Shimoakka Hatchery.

REFERENCES

ASC. (2017). *ASC Salmon Standard. V1.1.* Retrieved from https://www.asc-aqua.org/

FAO. (2009). *Guidelines for the ecolabelling of fish and fishery products from marine capture fisheries.* Retrieved from http://www.fao.org/in-action/globefish/publications/details-publication/en/c/346098/

FAO. (2018a). Global capture production: Quantities (1950-2016). *Download dataset for FAO FishStat J.*

FAO. (2018b). Global aquaculture production: Quantities (1950-2016). *Download dataset for FAO FishStat J.*

Grau, E. G., Dickhoff, W. W., Bern, H. A., & Folmar, L. C. (1981). Lunar phasing of the thyroxine preparatory to seaward migration of salmonid fish. *Science, 211*(4482), 607–609. doi:10.1126cience.7455703 PMID:7455703

Hiratsuka, S., Koizumi, K., Ooba, T., & Yokogoshi, H. (2009). Effects of dietary docosahexaenoic acid connecting phospholipids on the learning ability and fatty acid composition of the brain. *Journal of Nutritional Science and Vitaminology, 55*(4), 374–380. doi:10.3177/jnsv.55.374 PMID:19763040

Houston, R. D., & Macqueen, D. J. (2019). Atlantic salmon (Salmo salar L.) genetics in the 21st century: Taking leaps forward in aquaculture and biological understanding. *Animal Genetics, 50*(1), 3–14. doi:10.1111/age.12748 PMID:30426521

Kaeriyama, M., Seo, H., & Qin, Y. (2014). Effects of global warming on the life history and population dynamics of Japanese chum salmon. *Fisheries Science, 80*(2), 251–260. doi:10.100712562-013-0693-7

Kitada, S. (2014). Japanese chum salmon stock enhancement: Current perspective and future challenges. *Fisheries Science, 80*(2), 237–249. doi:10.100712562-013-0692-8

Lhorente, J. P., Aranneda, M., Neira, R., & Yanez, J. M. (2019, May). Advances in genetic improvement for salmon and trout aquaculture: The Chilean situation and prospects. *Reviews in Aquaculture, 11*(2), 340–353. doi:10.1111/raq.12335

Mizuno, S., Hatakeyama, M., Nakajima, M., Naito, K., Koyama, T., Saneyoshi, H., ... Ueda, H. (2010b). Relationship between rearing condition and health in chum salmon (*Oncorhynchus keta*) fry. *Suisan Zoshoku, 58*, 529–531. doi:10.11233/aquaculturesci.58.529

Mizuno, S., Nakajima, M., Naito, K., Koyama, T., Saneyoshi, H., Kobayashi, M., ... Ueda, H. (2010a). Physiological impacts on high rearing density on chum salmon *Oncorhynchus keta* fry. *Suisan Zoshoku, 58*, 387–399. doi:10.11233/aquaculturesci.58.387

Nakamori, T., Maekawa, F., Sato, K., Tanaka, K., & Ohki-Hamazaki, H. (2013). Neural basis of imprinting behavior in chicks. *Development, Growth & Differentiation, 55*(1), 198–206. doi:10.1111/dgd.12028 PMID:23294362

Naylor, R. L., Goldburg, R. J., Mooney, H., Beveridge, M., Clay, J., Folke, C., ... Williams, M. (1998). Nature's subsides to shrimp and salmon farming. *Science, 282*(5390), 883–884. doi:10.1126cience.282.5390.883

Otomo, T., Shimizu, Y., & Takahashi, N. (2006). Development of stock enhancement technique for *Oncorhynchus masou* [in Japanese with English summary]. *Bulletin of Iwate Prefectural Fisheries Technology Center, 6*, 7–13.

Pauly, D., & Zeller, D. (2016). Catch reconstructions reveal that global marine fisheries catches are higher than reported and declining. *Nature Communications, 7*(1), 19244. doi:10.1038/ncomms10244 PMID:26784963

Quinones, R. A., Fuentes, M., Montes, R. M., Soto, D., & Leon-Munoz, J. (2019). Environmental issues in Chilean salmon farming: A review. *Reviews in Aquaculture, 11*(2), 375–402. doi:10.1111/raq.12337

Seo, H., Kudo, H., & Kaeriyama, M. (2011). Long-term climate-related changes in somatic growth and population dynamics of Hokkaido chum salmon. *Environmental Biology of Fishes, 90*(2), 131–142. doi:10.100710641-010-9725-7

Taranger, F. L., Karlsen, Ø., Bannister, R. J., Glover, K. A., Husa, V., Karlsbakk, E., ... Svasand, T. (2015). Risk assessment of the environmental impact of Norwegian Atlantic salmon farming. *ICES Journal of Marine Science, 72*(3), 997–1021. doi:10.1093/icesjms/fsu132

Ueda, H. (2015). The current propagation system and physiological studies of Japanese chum salmon. *Fisheries Science*, *81*(2), 219–228. doi:10.100712562-014-0847-2

Urawa, H. (2004). Stock identification studies of high seas salmon in Japan: A review and future plan. *North Pacific Anadromous Fish Technical Report*, *5*, 9–10.

Yamaguchi, S., Aoki, N., Kitajima, T., Iikubo, E., Katagiri, S., Matsushima, T., & Homma, K. (2012). Thyroid hormone determines the start of the sensitive period of imprinting and primes later learning. *Nature Communications*, *3*(1), 1081. doi:10.1038/ncomms2088 PMID:23011135

Yamauchi, K., Ban, M., Kasahara, N., Izumi, T., Kojima, H., & Harako, T. (1985). Physiological and behavioral changes occurring during smoltification in the masu salmon, *Oncorhynchus masou. Aquaculture (Amsterdam, Netherlands)*, *45*(1-4), 227–236. doi:10.1016/0044-8486(85)90272-8

Related Readings

To continue IGI Global's long-standing tradition of advancing innovation through emerging research, please find below a compiled list of recommended IGI Global book chapters and journal articles in the areas of water resources, climate change, and fishery. These related readings will provide additional information and guidance to further enrich your knowledge and assist you with your own research.

Adler, M. (2015). Floods Monitoring. In C. Maftei (Ed.), *Extreme Weather and Impacts of Climate Change on Water Resources in the Dobrogea Region* (pp. 312–344). Hershey, PA: IGI Global. doi:10.4018/978-1-4666-8438-6.ch011

Afzal, S. (2016). Implementation of Flooding Free Routing in Smart Grid: VCP Routing in Smart Gird. In A. Ahmad & N. Hassan (Eds.), *Smart Grid as a Solution for Renewable and Efficient Energy* (pp. 298–322). Hershey, PA: IGI Global. doi:10.4018/978-1-5225-0072-8.ch013

Ahmad, S., Ahmad, A., & Yaqub, R. (2016). Optimized Energy Consumption and Demand Side Management in Smart Grid. In A. Ahmad & N. Hassan (Eds.), *Smart Grid as a Solution for Renewable and Efficient Energy* (pp. 1–25). Hershey, PA: IGI Global. doi:10.4018/978-1-5225-0072-8.ch001

Arbaiy, N., Watada, J., & Lin, P. (2016). Fuzzy Random Regression-Based Modeling in Uncertain Environment. In P. Vasant & N. Voropai (Eds.), *Sustaining Power Resources through Energy Optimization and Engineering* (pp. 127–146). Hershey, PA: IGI Global. doi:10.4018/978-1-4666-9755-3.ch006

Arhin, A. (2016). Improving Sustainability of the Environment in a Changing Climate: Can REDD+ Rise to the Challenge? In S. Dinda (Ed.), *Handbook of Research on Climate Change Impact on Health and Environmental Sustainability* (pp. 327–346). Hershey, PA: IGI Global. doi:10.4018/978-1-4666-8814-8.ch016

Bahinipati, C. S., Patnaik, U., & Viswanathan, P. K. (2016). What Causes Economic Losses from Natural Disasters in India? In S. Dinda (Ed.), *Handbook of Research on Climate Change Impact on Health and Environmental Sustainability* (pp. 157–175). Hershey, PA: IGI Global. doi:10.4018/978-1-4666-8814-8.ch008

Banerjee, S. (2016). Social Innovation: A Theoretical Approach in Intertwining Climate Change with Social Innovation. In S. Dinda (Ed.), *Handbook of Research on Climate Change Impact on Health and Environmental Sustainability* (pp. 593–618). Hershey, PA: IGI Global. doi:10.4018/978-1-4666-8814-8.ch029

Barakabitze, A. A., Fue, K. G., Kitindi, E. J., & Sanga, C. A. (2016). Developing a Framework for Next Generation Integrated Agro Food-Advisory Systems in Developing Countries. *International Journal of Information Communication Technologies and Human Development*, *8*(4), 13–31. doi:10.4018/IJICTHD.2016100102

Basu, J. P. (2016). Coastal Poverty, Resource-Dependent Livelihood, Climate Change, and Adaptation: An Empirical Study in Indian Coastal Sunderbans. In S. Dinda (Ed.), *Handbook of Research on Climate Change Impact on Health and Environmental Sustainability* (pp. 441–454). Hershey, PA: IGI Global. doi:10.4018/978-1-4666-8814-8.ch022

Bekele, I., & Ganpat, W. (2015). Education, Extension, and Training for Climate Change. In W. Ganpat & W. Isaac (Eds.), *Impacts of Climate Change on Food Security in Small Island Developing States* (pp. 361–388). Hershey, PA: IGI Global. doi:10.4018/978-1-4666-6501-9.ch012

Bhaskar, A., Rao, G. B., & Vencatesan, J. (2017). Characterization and Management Concerns of Water Resources around Pallikaranai Marsh, South Chennai. In P. Rao & Y. Patil (Eds.), *Reconsidering the Impact of Climate Change on Global Water Supply, Use, and Management* (pp. 102–121). Hershey, PA: IGI Global. doi:10.4018/978-1-5225-1046-8.ch007

Bhatt, R. (2017). Zero Tillage for Mitigating Global Warming Consequences and Improving Livelihoods in South Asia. In W. Ganpat & W. Isaac (Eds.), *Environmental Sustainability and Climate Change Adaptation Strategies* (pp. 126–161). Hershey, PA: IGI Global. doi:10.4018/978-1-5225-1607-1.ch005

Bit, J., & Banerjee, S. (2016). Sustainable Forest Use and India's Economic Growth: A Structural Decomposition Analysis of Direct Forest Intensity. In S. Dinda (Ed.), *Handbook of Research on Climate Change Impact on Health and Environmental Sustainability* (pp. 306–326). Hershey, PA: IGI Global. doi:10.4018/978-1-4666-8814-8.ch015

Boonkerd, K. (2017). Development and Modification of Natural Rubber for Advanced Application. In T. Kobayashi (Ed.), *Applied Environmental Materials Science for Sustainability* (pp. 44–76). Hershey, PA: IGI Global. doi:10.4018/978-1-5225-1971-3.ch003

Boonmahitthisud, A. (2017). Natural Rubber and Rubber Blend Nanocomposites: Reinforcement of Natural Rubber with Polymer-Encapsulated Inorganic Nanohybrid Particles. In T. Kobayashi (Ed.), *Applied Environmental Materials Science for Sustainability* (pp. 77–105). Hershey, PA: IGI Global. doi:10.4018/978-1-5225-1971-3.ch004

Bostanci, S. H., & Albayrak, A. N. (2017). The Role of Eco-Municipalities in Climate Change for a Sustainable Future. In W. Ganpat & W. Isaac (Eds.), *Environmental Sustainability and Climate Change Adaptation Strategies* (pp. 213–231). Hershey, PA: IGI Global. doi:10.4018/978-1-5225-1607-1.ch008

Buta, C., Omer, I., & Andronic, A. (2015). Hydrological Risk Phenomena and Flood Analysis: Study Case – Taita Catchment, Romania. In C. Maftei (Ed.), *Extreme Weather and Impacts of Climate Change on Water Resources in the Dobrogea Region* (pp. 284–311). Hershey, PA: IGI Global. doi:10.4018/978-1-4666-8438-6.ch010

Carrillo, K. L., & Kobayashi, T. (2017). Natural Material Source of Bagasse Cellulose and Their Application to Hydrogel Films. In T. Kobayashi (Ed.), *Applied Environmental Materials Science for Sustainability* (pp. 19–43). Hershey, PA: IGI Global. doi:10.4018/978-1-5225-1971-3.ch002

Cazacu, G. B. (2015). Dobrogea Geology. In C. Maftei (Ed.), *Extreme Weather and Impacts of Climate Change on Water Resources in the Dobrogea Region* (pp. 73–118). Hershey, PA: IGI Global. doi:10.4018/978-1-4666-8438-6.ch004

Chai-Ittipornwong, T. (2017). Participation Framework to Sustainability: The Undercurrents in Bottled-Water Production and Consumption. In P. Rao & Y. Patil (Eds.), *Reconsidering the Impact of Climate Change on Global Water Supply, Use, and Management* (pp. 272–293). Hershey, PA: IGI Global. doi:10.4018/978-1-5225-1046-8.ch015

Chatterjee, T., & Dinda, S. (2016). Climate Change, Human Health and Some Economic Issues. In S. Dinda (Ed.), *Handbook of Research on Climate Change Impact on Health and Environmental Sustainability* (pp. 26–41). Hershey, PA: IGI Global. doi:10.4018/978-1-4666-8814-8.ch002

Chaudhary, N., & Pisolkar, Y. (2017). Issues, Concerns, and Local Stakes: Future of Water Resources in Coastal Villages of Devbag and Tarkarli, Coastal Maharashtra, India. In P. Rao & Y. Patil (Eds.), *Reconsidering the Impact of Climate Change on Global Water Supply, Use, and Management* (pp. 50–69). Hershey, PA: IGI Global. doi:10.4018/978-1-5225-1046-8.ch004

Chelcea, S., Ionita, M., & Adler, M. (2015). Identification of Dry Periods in the Dobrogea Region. In C. Maftei (Ed.), *Extreme Weather and Impacts of Climate Change on Water Resources in the Dobrogea Region* (pp. 52–72). Hershey, PA: IGI Global. doi:10.4018/978-1-4666-8438-6.ch003

Cohen, J. E., Clarke-Harris, D. O., Khan, A., & Isaac, W. P. (2015). Sustainable Management of Invasive Species for Small Island Developing States under Changing Climates. In W. Ganpat & W. Isaac (Eds.), *Impacts of Climate Change on Food Security in Small Island Developing States* (pp. 312–360). Hershey, PA: IGI Global. doi:10.4018/978-1-4666-6501-9.ch011

Das, S. (2016). Health Impact of Water-Related Diseases in Developing Countries on Account of Climate Change – A Systematic Review: A Study in Regard to South Asian Countries. In S. Dinda (Ed.), *Handbook of Research on Climate Change Impact on Health and Environmental Sustainability* (pp. 42–60). Hershey, PA: IGI Global. doi:10.4018/978-1-4666-8814-8.ch003

David, I., Beilicci, E., & Beilicci, R. (2015). Basics for Hydraulic Modelling of Flood Runoff Using Advanced Hydroinformatic Tools. In C. Maftei (Ed.), *Extreme Weather and Impacts of Climate Change on Water Resources in the Dobrogea Region* (pp. 205–239). Hershey, PA: IGI Global. doi:10.4018/978-1-4666-8438-6.ch008

Deenapanray, P. N., & Ramma, I. (2015). Adaptations to Climate Change and Climate Variability in the Agriculture Sector in Mauritius: Lessons from a Technical Needs Assessment. In W. Ganpat & W. Isaac (Eds.), *Impacts of Climate Change on Food Security in Small Island Developing States* (pp. 130–165). Hershey, PA: IGI Global. doi:10.4018/978-1-4666-6501-9.ch005

Dinda, S. (2016). Adaptation to Climate Change for Sustainable Development: A Survey. In S. Dinda (Ed.), *Handbook of Research on Climate Change Impact on Health and Environmental Sustainability* (pp. 363–391). Hershey, PA: IGI Global. doi:10.4018/978-1-4666-8814-8.ch018

Dinda, S. (2016). Climate Change, Trade Competitiveness, and Opportunity for Climate Friendly Goods in SAARC and Asia Pacific Regions. In S. Dinda (Ed.), *Handbook of Research on Climate Change Impact on Health and Environmental Sustainability* (pp. 515–536). Hershey, PA: IGI Global. doi:10.4018/978-1-4666-8814-8.ch026

Dongol, D., Bollin, E., & Feldmann, T. (2016). Battery Management Based on Predictive Control and Demand-Side Management: Smart Integration of Renewable Energy Sources. In A. Ahmad & N. Hassan (Eds.), *Smart Grid as a Solution for Renewable and Efficient Energy* (pp. 149–180). Hershey, PA: IGI Global. doi:10.4018/978-1-5225-0072-8.ch007

Drăguşin, D. (2015). Drought Effects on Groundwater in Dobrogea Plateau. In C. Maftei (Ed.), *Extreme Weather and Impacts of Climate Change on Water Resources in the Dobrogea Region* (pp. 119–144). Hershey, PA: IGI Global. doi:10.4018/978-1-4666-8438-6.ch005

Elsayed, A. M., Dakkama, H. J., Mahmoud, S., Al-Dadah, R., & Kaialy, W. (2017). Sustainable Cooling Research Using Activated Carbon Adsorbents and Their Environmental Impact. In T. Kobayashi (Ed.), *Applied Environmental Materials Science for Sustainability* (pp. 186–221). Hershey, PA: IGI Global. doi:10.4018/978-1-5225-1971-3.ch009

Emmanuel, M., Muasa, L., Chen, C., Mutisya, F., & Avtar, R. (2016). Impact of Rapid Urbanization and Climate Change on Agricultural Productivity in Africa: Climate Change Policies in the Agricultural Sector. In S. Dinda (Ed.), *Handbook of Research on Climate Change Impact on Health and Environmental Sustainability* (pp. 416–426). Hershey, PA: IGI Global. doi:10.4018/978-1-4666-8814-8.ch020

Emran, A., Rob, M. A., & Kabir, M. H. (2017). Coastline Change and Erosion-Accretion Evolution of the Sandwip Island, Bangladesh. *International Journal of Applied Geospatial Research, 8*(2), 33–44. doi:10.4018/IJAGR.2017040103

Eudoxie, G., & Roopnarine, R. (2017). Climate Change Adaptation and Disaster Risk Management in the Caribbean. In W. Ganpat & W. Isaac (Eds.), *Environmental Sustainability and Climate Change Adaptation Strategies* (pp. 97–125). Hershey, PA: IGI Global. doi:10.4018/978-1-5225-1607-1.ch004

Eudoxie, G. D., & Wuddivira, M. (2015). Soil, Water, and Agricultural Adaptations. In W. Ganpat & W. Isaac (Eds.), *Impacts of Climate Change on Food Security in Small Island Developing States* (pp. 255–279). Hershey, PA: IGI Global. doi:10.4018/978-1-4666-6501-9.ch009

Fernández, F. J., Jiménez, A. D., Manzano, F. S., & Márquez, J. M. (2017). An Energy Management Strategy and Fuel Cell Configuration Proposal for a Hybrid Renewable System with Hydrogen Backup. *International Journal of Energy Optimization and Engineering, 6*(1), 1–22. doi:10.4018/IJEOE.2017010101

Fernando, Y., & Wah, W. X. (2016). Moving forward a Parsimonious Model of Eco-Innovation: Results from a Content Analysis. In S. Dinda (Ed.), *Handbook of Research on Climate Change Impact on Health and Environmental Sustainability* (pp. 619–631). Hershey, PA: IGI Global. doi:10.4018/978-1-4666-8814-8.ch030

Garrick, T. A., & Liburd, O. E. (2017). Impact of Climate Change on a Key Agricultural Pest: Thrips. In W. Ganpat & W. Isaac (Eds.), *Environmental Sustainability and Climate Change Adaptation Strategies* (pp. 232–254). Hershey, PA: IGI Global. doi:10.4018/978-1-5225-1607-1.ch009

Goundar, S., & Appana, S. (2017). Mainstreaming Development Policies for Climate Change in Fiji: A Policy Gap Analysis and the Role of ICTs. In W. Ganpat & W. Isaac (Eds.), *Environmental Sustainability and Climate Change Adaptation Strategies* (pp. 1–31). Hershey, PA: IGI Global. doi:10.4018/978-1-5225-1607-1.ch001

Gupta, A. C. (2016). Bioeconomic Fishery Management: Changing Paradigms towards Eco-System Based Management. In S. Dinda (Ed.), *Handbook of Research on Climate Change Impact on Health and Environmental Sustainability* (pp. 261–281). Hershey, PA: IGI Global. doi:10.4018/978-1-4666-8814-8.ch013

Hiremath, R., Kumar, B., Bansode, S. S., Nulkar, G., Patil, S. S., & Murali, J. (2017). Industrial Wastewater Management in the Context of Climate Change Adaptation in Selected Cities of India: A Business Approach. In P. Rao & Y. Patil (Eds.), *Reconsidering the Impact of Climate Change on Global Water Supply, Use, and Management* (pp. 294–313). Hershey, PA: IGI Global. doi:10.4018/978-1-5225-1046-8.ch016

Huyen, P. T. (2017). Clay Minerals Converted to Porous Materials and Their Application: Challenge and Perspective. In T. Kobayashi (Ed.), *Applied Environmental Materials Science for Sustainability* (pp. 141–164). Hershey, PA: IGI Global. doi:10.4018/978-1-5225-1971-3.ch007

Iese, V., Maeke, J., Holland, E., Wairiu, M., & Naidu, S. (2015). Farming Adaptations to the Impacts of Climate Change and Extreme Events in Pacific Island Countries: Case Study of Bellona Atoll, Solomon Islands. In W. Ganpat & W. Isaac (Eds.), *Impacts of Climate Change on Food Security in Small Island Developing States* (pp. 166–194). Hershey, PA: IGI Global. doi:10.4018/978-1-4666-6501-9.ch006

Ikematsu, S., Tada, I., & Nagasaki, Y. (2017). Analysis of Lipids Produced by Microalgae Isolated from the Area around Okinawa, Japan. In T. Kobayashi (Ed.), *Applied Environmental Materials Science for Sustainability* (pp. 222–233). Hershey, PA: IGI Global. doi:10.4018/978-1-5225-1971-3.ch010

Inogwabini, B. (2017). Congo Basin's Shrinking Watersheds: Potential Consequences on Local Communities. In P. Rao & Y. Patil (Eds.), *Reconsidering the Impact of Climate Change on Global Water Supply, Use, and Management* (pp. 211–226). Hershey, PA: IGI Global. doi:10.4018/978-1-5225-1046-8.ch012

Ionita, M., & Chelcea, S. (2015). Spatio-Temporal Variability of Seasonal Drought over the Dobrogea Region. In C. Maftei (Ed.), *Extreme Weather and Impacts of Climate Change on Water Resources in the Dobrogea Region* (pp. 17–51). Hershey, PA: IGI Global. doi:10.4018/978-1-4666-8438-6.ch002

Issakhov, A. (2016). Mathematical Modelling of the Thermal Process in the Aquatic Environment with Considering the Hydrometeorological Condition at the Reservoir-Cooler by Using Parallel Technologies. In P. Vasant & N. Voropai (Eds.), *Sustaining Power Resources through Energy Optimization and Engineering* (pp. 227–243). Hershey, PA: IGI Global. doi:10.4018/978-1-4666-9755-3.ch010

Javid, T. (2016). Geographic Information System for the Smart Grid. In A. Ahmad & N. Hassan (Eds.), *Smart Grid as a Solution for Renewable and Efficient Energy* (pp. 344–362). Hershey, PA: IGI Global. doi:10.4018/978-1-5225-0072-8.ch015

Juma, D. W., Reuben, M., Wang, H., & Li, F. (2017). Adaptive Coevolution: Realigning the Water Governance Regime to the Changing Climate. In P. Rao & Y. Patil (Eds.), *Reconsidering the Impact of Climate Change on Global Water Supply, Use, and Management* (pp. 314–325). Hershey, PA: IGI Global. doi:10.4018/978-1-5225-1046-8.ch017

Kais, S. M. (2017). Climate Change: Vulnerability and Resilience in Commercial Shrimp Aquaculture in Bangladesh. In W. Ganpat & W. Isaac (Eds.), *Environmental Sustainability and Climate Change Adaptation Strategies* (pp. 162–187). Hershey, PA: IGI Global. doi:10.4018/978-1-5225-1607-1.ch006

Kamboj, V. K., & Bath, S. K. (2016). Scope of Biogeography-Based Optimization for Economic Load Dispatch and Multi-Objective Unit Commitment Problem. In P. Vasant & N. Voropai (Eds.), *Sustaining Power Resources through Energy Optimization and Engineering* (pp. 360–389). Hershey, PA: IGI Global. doi:10.4018/978-1-4666-9755-3.ch015

Karmaoui, A. (2016). Environmental Vulnerability to Climate Change in Mediterranean Basin: Socio-Ecological Interactions between North and South. In S. Dinda (Ed.), *Handbook of Research on Climate Change Impact on Health and Environmental Sustainability* (pp. 105–138). Hershey, PA: IGI Global. doi:10.4018/978-1-4666-8814-8.ch006

Khai, H. V. (2016). Assessing Urban Residents' Willingness to Pay for Preserving the Biodiversity of Swamp Forest. In S. Dinda (Ed.), *Handbook of Research on Climate Change Impact on Health and Environmental Sustainability* (pp. 283–305). Hershey, PA: IGI Global. doi:10.4018/978-1-4666-8814-8.ch014

Khalid, S. (2016). Application of Adaptive Tabu Search Algorithm in Hybrid Power Filter and Shunt Active Power Filters: Application of ATS Algorithm in HPF and APF. In P. Vasant & N. Voropai (Eds.), *Sustaining Power Resources through Energy Optimization and Engineering* (pp. 276–308). Hershey, PA: IGI Global. doi:10.4018/978-1-4666-9755-3.ch012

Khan, S. S. (2016). Modeling and Operating Strategies of Micro-Grids for Renewable Energy Communities. In A. Ahmad & N. Hassan (Eds.), *Smart Grid as a Solution for Renewable and Efficient Energy* (pp. 97–122). Hershey, PA: IGI Global. doi:10.4018/978-1-5225-0072-8.ch005

Khanna, B. K. (2017). Indian National Strategy for Climate Change Adaptation and Mitigation. In W. Ganpat & W. Isaac (Eds.), *Environmental Sustainability and Climate Change Adaptation Strategies* (pp. 32–63). Hershey, PA: IGI Global. doi:10.4018/978-1-5225-1607-1.ch002

Khanna, B. K. (2017). Vulnerability of the Lakshadweep Coral Islands in India and Strategies for Mitigating Climate Change Impacts. In W. Ganpat & W. Isaac (Eds.), *Environmental Sustainability and Climate Change Adaptation Strategies* (pp. 64–96). Hershey, PA: IGI Global. doi:10.4018/978-1-5225-1607-1.ch003

Kobayashi, T. (2017). Introduction of Environmental Materials. In T. Kobayashi (Ed.), *Applied Environmental Materials Science for Sustainability* (pp. 1–18). Hershey, PA: IGI Global. doi:10.4018/978-1-5225-1971-3.ch001

Kumar, C. P. (2016). Impact of Climate Change on Groundwater Resources. In S. Dinda (Ed.), *Handbook of Research on Climate Change Impact on Health and Environmental Sustainability* (pp. 196–221). Hershey, PA: IGI Global. doi:10.4018/978-1-4666-8814-8.ch010

Kumar, K. V., Kumar, S. S., Selvakumar, A. I., & Kumar, R. S. (2016). Recent Techniques to Identify the Stator Fault Diagnosis in Three Phase Induction Motor. In P. Vasant & N. Voropai (Eds.), *Sustaining Power Resources through Energy Optimization and Engineering* (pp. 309–325). Hershey, PA: IGI Global. doi:10.4018/978-1-4666-9755-3.ch013

Kumar, M. (2016). Environmental Sustainability in the Fashion Supply Chain in India. *International Journal of Social Ecology and Sustainable Development*, *7*(3), 1–33. doi:10.4018/IJSESD.2016070101

Kumar, M. (2017). Economic Evaluation of Solar Cooling Schemes. *International Journal of Energy Optimization and Engineering*, *6*(1), 23–48. doi:10.4018/IJEOE.2017010102

Kumar, R., Rao, P., & Areendran, G. (2017). Understanding Glacial Retreat in the Indian Himalaya: Historical Trends and Field Studies from a Large Glacier. In P. Rao & Y. Patil (Eds.), *Reconsidering the Impact of Climate Change on Global Water Supply, Use, and Management* (pp. 33–49). Hershey, PA: IGI Global. doi:10.4018/978-1-5225-1046-8.ch003

Kumari, S., & Patil, Y. (2017). Achieving Climate Smart Agriculture with a Sustainable Use of Water: A Conceptual Framework for Sustaining the Use of Water for Agriculture in the Era of Climate Change. In P. Rao & Y. Patil (Eds.), *Reconsidering the Impact of Climate Change on Global Water Supply, Use, and Management* (pp. 122–143). Hershey, PA: IGI Global. doi:10.4018/978-1-5225-1046-8.ch008

Kumari, S., Patil, Y., & Rao, P. (2017). An Approach to Sustainable Watershed Management: Case Studies on Enhancing Sustainability with Challenges of Water in Western Maharashtra. In P. Rao & Y. Patil (Eds.), *Reconsidering the Impact of Climate Change on Global Water Supply, Use, and Management* (pp. 252–271). Hershey, PA: IGI Global. doi:10.4018/978-1-5225-1046-8.ch014

Lallo, C. H., Smalling, S., Facey, A., & Hughes, M. (2017). The Impact of Climate Change on Small Ruminant Performance in Caribbean Communities. In W. Ganpat & W. Isaac (Eds.), *Environmental Sustainability and Climate Change Adaptation Strategies* (pp. 296–321). Hershey, PA: IGI Global. doi:10.4018/978-1-5225-1607-1.ch011

Lawrence, J., Simpson, L., & Piggott, A. (2015). Protected Agriculture: A Climate Change Adaptation for Food and Nutrition Security. In W. Ganpat & W. Isaac (Eds.), *Impacts of Climate Change on Food Security in Small Island Developing States* (pp. 196–220). Hershey, PA: IGI Global. doi:10.4018/978-1-4666-6501-9.ch007

Li, K., & Kobayashi, T. (2017). Ionic Liquids and Poly (Ionic Liquid)s Used as Green Solvent and Ultrasound Responded Materials. In T. Kobayashi (Ed.), *Applied Environmental Materials Science for Sustainability* (pp. 327–346). Hershey, PA: IGI Global. doi:10.4018/978-1-5225-1971-3.ch015

Liapakis, A., Costopoulou, C., Tsiligiridis, T., & Sideridis, A. (2017). Studying Corporate Social Responsibility Activities in the Agri-Food Sector: The Greek Case. *International Journal of Agricultural and Environmental Information Systems*, *8*(1), 1–13. doi:10.4018/IJAEIS.2017010101

Loi, N. K., Huyen, N. T., Tu, L. H., Tram, V. N., Liem, N. D., Dat, N. L., ... Minh, D. N. (2017). Sustainable Land Use and Watershed Management in Response to Climate Change Impacts: Case Study in Srepok Watershed, Central Highland of Vietnam. In W. Ganpat & W. Isaac (Eds.), *Environmental Sustainability and Climate Change Adaptation Strategies* (pp. 255–295). Hershey, PA: IGI Global. doi:10.4018/978-1-5225-1607-1.ch010

Londhe, S. (2016). Impact of Climate Change on Agriculture and Food Security. *International Journal of Disease Control and Containment for Sustainability*, *1*(1), 32–46. doi:10.4018/IJDCCS.2016010103

Londhe, S. (2017). Inter Linkages of Water, Climate, and Agriculture. In P. Rao & Y. Patil (Eds.), *Reconsidering the Impact of Climate Change on Global Water Supply, Use, and Management* (pp. 166–194). Hershey, PA: IGI Global. doi:10.4018/978-1-5225-1046-8.ch010

Londhe, S. L. (2016). Climate Change and Agriculture: Impacts, Adoption, and Mitigation. In S. Dinda (Ed.), *Handbook of Research on Climate Change Impact on Health and Environmental Sustainability* (pp. 393–415). Hershey, PA: IGI Global. doi:10.4018/978-1-4666-8814-8.ch019

Maftei, C., & Papatheodorou, K. (2015). Mathematical Models Used for Hydrological Floodplain Modeling. In C. Maftei (Ed.), *Extreme Weather and Impacts of Climate Change on Water Resources in the Dobrogea Region* (pp. 240–283). Hershey, PA: IGI Global. doi:10.4018/978-1-4666-8438-6.ch009

Maharaj, R., Singh-Ackbarali, D., & Sankat, C. K. (2015). Postharvest Management Strategies. In W. Ganpat & W. Isaac (Eds.), *Impacts of Climate Change on Food Security in Small Island Developing States* (pp. 221–254). Hershey, PA: IGI Global. doi:10.4018/978-1-4666-6501-9.ch008

Mahboob, Q. (2016). Identification of Reliability Critical Items in Large and Complex Rail Electrical Networks. In A. Ahmad & N. Hassan (Eds.), *Smart Grid as a Solution for Renewable and Efficient Energy* (pp. 226–248). Hershey, PA: IGI Global. doi:10.4018/978-1-5225-0072-8.ch010

Mahmood, W. A., & Azarian, M. H. (2017). Inorganic-Organic Composite Materials from Liquid Natural Rubber and Epoxidised Natural Rubber Derivatives: Prospects and Applications. In T. Kobayashi (Ed.), *Applied Environmental Materials Science for Sustainability* (pp. 128–140). Hershey, PA: IGI Global. doi:10.4018/978-1-5225-1971-3.ch006

Majumder, J., Shah, P., & Kumar, S. (2016). Heat Stress Vulnerability among Indian Workmen. In S. Dinda (Ed.), *Handbook of Research on Climate Change Impact on Health and Environmental Sustainability* (pp. 61–80). Hershey, PA: IGI Global. doi:10.4018/978-1-4666-8814-8.ch004

Maximay, S. (2015). The Caribbean's Response to Climate Change Impacts. In W. Ganpat & W. Isaac (Eds.), *Impacts of Climate Change on Food Security in Small Island Developing States* (pp. 33–66). Hershey, PA: IGI Global. doi:10.4018/978-1-4666-6501-9.ch002

Meldrum, H. M., Szymanski, D., Oches, E. A., & Davis, P. T. (2016). A Picture is Worth a Thousand Words: Commentary of Broadcast Meteorologists on the Visual Presentation of Climate Change. *International Journal of Social Ecology and Sustainable Development*, *7*(4), 1–16. doi:10.4018/IJSESD.2016100101

Mili, B., Barua, A., & Katyaini, S. (2016). Climate Change and Adaptation through the Lens of Capability Approach: A Case Study from Darjeeling, Eastern Himalaya. In S. Dinda (Ed.), *Handbook of Research on Climate Change Impact on Health and Environmental Sustainability* (pp. 455–469). Hershey, PA: IGI Global. doi:10.4018/978-1-4666-8814-8.ch023

Minea, G., Bandoc, G., & Neculau, G. (2015). Seasonal Statistical Variability of Precipitations in Dobrogea and Danube Delta. In C. Maftei (Ed.), *Extreme Weather and Impacts of Climate Change on Water Resources in the Dobrogea Region* (pp. 1–16). Hershey, PA: IGI Global. doi:10.4018/978-1-4666-8438-6.ch001

Minhas, D. M., & Hussain, S. (2016). Efficient Control Strategies to Optimize Electricity Cost and Consumer Satisfaction. In A. Ahmad & N. Hassan (Eds.), *Smart Grid as a Solution for Renewable and Efficient Energy* (pp. 69–96). Hershey, PA: IGI Global. doi:10.4018/978-1-5225-0072-8.ch004

Mohapatra, M., & Baladhandautham, C. B. (2016). Implementation of Improved Control Strategy of DC-AC Converter using Delta-Sigma Modulator. In A. Ahmad & N. Hassan (Eds.), *Smart Grid as a Solution for Renewable and Efficient Energy* (pp. 249–272). Hershey, PA: IGI Global. doi:10.4018/978-1-5225-0072-8.ch011

Moustache, A. M. (2015). Adaptation to Impacts of Climate Change on the Food and Nutrition Security Status of a Small Island Developing State: The Case of the Republic of Seychelles. In W. Ganpat & W. Isaac (Eds.), *Impacts of Climate Change on Food Security in Small Island Developing States* (pp. 96–129). Hershey, PA: IGI Global. doi:10.4018/978-1-4666-6501-9.ch004

Mujere, N., & Moyce, W. (2017). Climate Change Impacts on Surface Water Quality. In W. Ganpat & W. Isaac (Eds.), *Environmental Sustainability and Climate Change Adaptation Strategies* (pp. 322–340). Hershey, PA: IGI Global. doi:10.4018/978-1-5225-1607-1.ch012

Mukherjee, S., & Chakraborty, D. (2016). Does Fiscal Policy Influence Per Capita CO2 Emission?: A Cross Country Empirical Analysis. In S. Dinda (Ed.), *Handbook of Research on Climate Change Impact on Health and Environmental Sustainability* (pp. 568–592). Hershey, PA: IGI Global. doi:10.4018/978-1-4666-8814-8.ch028

N'Yeurt, A. D., & Iese, V. (2015). Marine Plants as a Sustainable Source of Agri-Fertilizers for Small Island Developing States (SIDS). In W. Ganpat & W. Isaac (Eds.), *Impacts of Climate Change on Food Security in Small Island Developing States* (pp. 280–311). Hershey, PA: IGI Global. doi:10.4018/978-1-4666-6501-9.ch010

Nersesian, R. L., & Strang, K. D. (2017). Feasibility Approaches to Reduce the Unreliability of Gas, Nuclear, Coal, Solar and Wind Electricity Production. *International Journal of Risk and Contingency Management*, *6*(1), 54–69. doi:10.4018/IJRCM.2017010104

Ngubevana, L. (2017). Sustainable Development Dilemmas of Biofuels Research and Production: A Snapshot in South Africa. *International Journal of Energy Optimization and Engineering*, *6*(2), 24–41. doi:10.4018/IJEOE.2017040102

Nyangon, J., Alabbas, N., & Agbemabiese, L. (2017). Entangled Systems at the Energy-Water-Food Nexus: Challenges and Opportunities. In P. Rao & Y. Patil (Eds.), *Reconsidering the Impact of Climate Change on Global Water Supply, Use, and Management* (pp. 144–165). Hershey, PA: IGI Global. doi:10.4018/978-1-5225-1046-8.ch009

Onutai, S., Jiemsirilers, S., & Kobayashi, T. (2017). Geopolymer Sourced with Fly Ash and Industrial Aluminum Waste for Sustainable Materials. In T. Kobayashi (Ed.), *Applied Environmental Materials Science for Sustainability* (pp. 165–185). Hershey, PA: IGI Global. doi:10.4018/978-1-5225-1971-3.ch008

Osmani, A. R. (2016). Greenhouse Gas Mitigation through Energy Efficiency: Perform, Achieve, and Trade (PAT) – India's Emission Trading Scheme. In S. Dinda (Ed.), *Handbook of Research on Climate Change Impact on Health and Environmental Sustainability* (pp. 537–566). Hershey, PA: IGI Global. doi:10.4018/978-1-4666-8814-8.ch027

Osmani, A. R. (2017). Tipaimukh Multipurpose Hydroelectric Project: A Policy Perspective – Indo-Bangla Priorities, Indigenous Peoples' Rights, and Environmental Concerns. In P. Rao & Y. Patil (Eds.), *Reconsidering the Impact of Climate Change on Global Water Supply, Use, and Management* (pp. 227–251). Hershey, PA: IGI Global. doi:10.4018/978-1-5225-1046-8.ch013

Ozpinar, A., & Ozil, E. (2016). Smart Grid and Demand Side Management: Application of Metaheuristic and Artificial Intelligence Algorithms. In A. Ahmad & N. Hassan (Eds.), *Smart Grid as a Solution for Renewable and Efficient Energy* (pp. 49–68). Hershey, PA: IGI Global. doi:10.4018/978-1-5225-0072-8.ch003

Padigala, B. S. (2017). Traditional Water Management System for Climate Change Adaptation in Mountain Ecosystems. In P. Rao & Y. Patil (Eds.), *Reconsidering the Impact of Climate Change on Global Water Supply, Use, and Management* (pp. 9–32). Hershey, PA: IGI Global. doi:10.4018/978-1-5225-1046-8.ch002

Page, T. (2017). A Feasibility Study in Energy Harvesting from Piezoelectric Keyboards. *International Journal of Energy Optimization and Engineering*, *6*(2), 1–23. doi:10.4018/IJEOE.2017040101

Paul, S., & Roy, P. K. (2016). A Novel Optimization Algorithm for Transient Stability Constrained Optimal Power Flow. In P. Vasant & N. Voropai (Eds.), *Sustaining Power Resources through Energy Optimization and Engineering* (pp. 147–176). Hershey, PA: IGI Global. doi:10.4018/978-1-4666-9755-3.ch007

Pieroni, A., & Iazeolla, G. (2016). Engineering QoS and Energy Saving in the Delivery of ICT Services. In P. Vasant & N. Voropai (Eds.), *Sustaining Power Resources through Energy Optimization and Engineering* (pp. 208–226). Hershey, PA: IGI Global. doi:10.4018/978-1-4666-9755-3.ch009

Polprasert, J., Ongsakul, W., & Dieu, V. N. (2016). Improved Pseudo-Gradient Search Particle Swarm Optimization for Optimal Power Flow Problem. In P. Vasant & N. Voropai (Eds.), *Sustaining Power Resources through Energy Optimization and Engineering* (pp. 177–207). Hershey, PA: IGI Global. doi:10.4018/978-1-4666-9755-3.ch008

Potiyaraj, P. (2017). Poly (Lactic Acid) Generated for Advanced Materials. In T. Kobayashi (Ed.), *Applied Environmental Materials Science for Sustainability* (pp. 106–127). Hershey, PA: IGI Global. doi:10.4018/978-1-5225-1971-3.ch005

Rale, V., & Tendulkar, P. (2017). Common Duckweeds as a Model System for Climate Change Impact Assessment. In P. Rao & Y. Patil (Eds.), *Reconsidering the Impact of Climate Change on Global Water Supply, Use, and Management* (pp. 364–372). Hershey, PA: IGI Global. doi:10.4018/978-1-5225-1046-8.ch019

Ramakrishna, A., & Bang, S. (2015). The Impacts of Climate Change on Food Security and Management in Papua New Guinea. In W. Ganpat & W. Isaac (Eds.), *Impacts of Climate Change on Food Security in Small Island Developing States* (pp. 67–95). Hershey, PA: IGI Global. doi:10.4018/978-1-4666-6501-9.ch003

Rani, S. (2016). Assessment of Annual, Monthly, and Seasonal Trends in the Long Term Rainfall of the Garhwal Himalayas. In S. Dinda (Ed.), *Handbook of Research on Climate Change Impact on Health and Environmental Sustainability* (pp. 222–241). Hershey, PA: IGI Global. doi:10.4018/978-1-4666-8814-8.ch011

Rao, P., & Patil, Y. (2017). Recent Trends, Issues, and Challenges in Water Resource Development and Global Climate Change. In P. Rao & Y. Patil (Eds.), *Reconsidering the Impact of Climate Change on Global Water Supply, Use, and Management* (pp. 1–8). Hershey, PA: IGI Global. doi:10.4018/978-1-5225-1046-8.ch001

Ray, S. (2016). A Framework for Understanding Adaptation by Manufacturing Industries. In S. Dinda (Ed.), *Handbook of Research on Climate Change Impact on Health and Environmental Sustainability* (pp. 471–481). Hershey, PA: IGI Global. doi:10.4018/978-1-4666-8814-8.ch024

Roberts, T. G., & Rodriguez, M. T. (2015). An Overview of Climate Change and Impacts on Food Security in Small Island Developing States. In W. Ganpat & W. Isaac (Eds.), *Impacts of Climate Change on Food Security in Small Island Developing States* (pp. 1–31). Hershey, PA: IGI Global. doi:10.4018/978-1-4666-6501-9.ch001

Roşu, L., & Macarov, L. I. (2015). Management of Drought and Floods in the Dobrogea Region. In C. Maftei (Ed.), *Extreme Weather and Impacts of Climate Change on Water Resources in the Dobrogea Region* (pp. 403–443). Hershey, PA: IGI Global. doi:10.4018/978-1-4666-8438-6.ch013

Roşu, L., & Zăgan, R. (2015). Management of Drought and Floods in Romania. In C. Maftei (Ed.), *Extreme Weather and Impacts of Climate Change on Water Resources in the Dobrogea Region* (pp. 345–402). Hershey, PA: IGI Global. doi:10.4018/978-1-4666-8438-6.ch012

Roy, J., Ghosh, D., Mukhopadhyay, K., & Ghosh, A. (2016). Exacerbating Health Risks in India due to Climate Change: Rethinking Approach to Health Service Provision. In S. Dinda (Ed.), *Handbook of Research on Climate Change Impact on Health and Environmental Sustainability* (pp. 1–25). Hershey, PA: IGI Global. doi:10.4018/978-1-4666-8814-8.ch001

Roy, P. K. (2016). A Novel Evolutionary Optimization Technique for Solving Optimal Reactive Power Dispatch Problems. In P. Vasant & N. Voropai (Eds.), *Sustaining Power Resources through Energy Optimization and Engineering* (pp. 244–275). Hershey, PA: IGI Global. doi:10.4018/978-1-4666-9755-3.ch011

Roy, P. K., Dutta, S., & Nandi, D. (2016). Optimal Reactive Power Dispatch Incorporating TCSC-TCPS Devices Using Different Evolutionary Optimization Techniques. In P. Vasant & N. Voropai (Eds.), *Sustaining Power Resources through Energy Optimization and Engineering* (pp. 326–359). Hershey, PA: IGI Global. doi:10.4018/978-1-4666-9755-3.ch014

Roy, P. K., & Ghosh, M. (2017). Combined Heat and Power Dispatch using Hybrid Genetic Algorithm and Biogeography-based Optimization. *International Journal of Energy Optimization and Engineering, 6*(1), 49–65. doi:10.4018/IJEOE.2017010103

Samanta, D. (2016). Lack of Land Tenure Security as Challenges to Sustainable Development: An Assessment in the Context of Bihar, India. In S. Dinda (Ed.), *Handbook of Research on Climate Change Impact on Health and Environmental Sustainability* (pp. 348–362). Hershey, PA: IGI Global. doi:10.4018/978-1-4666-8814-8.ch017

Sasirekha, S., & Swamynathan, S. (2017). Fuzzy Rule Based Environment Monitoring System for Weather Controlled Laboratories using Arduino. *International Journal of Intelligent Information Technologies, 13*(1), 50–66. doi:10.4018/IJIIT.2017010103

Satoh, M. (2017). Metal Ion Separation with Functional Adsorbents and Phytoremediation Used as Sustainable Technologies. In T. Kobayashi (Ed.), *Applied Environmental Materials Science for Sustainability* (pp. 284–312). Hershey, PA: IGI Global. doi:10.4018/978-1-5225-1971-3.ch013

Scorza, F. (2016). Towards Self Energy-Management and Sustainable Citizens' Engagement in Local Energy Efficiency Agenda. *International Journal of Agricultural and Environmental Information Systems, 7*(1), 44–53. doi:10.4018/IJAEIS.2016010103

Sen, S. K., & Pookayaporn, J. (2017). Role of Water-Energy-Waste Inter-Relatedness to Drive Sustainability amid Climate Concerns. In P. Rao & Y. Patil (Eds.), *Reconsidering the Impact of Climate Change on Global Water Supply, Use, and Management* (pp. 195–210). Hershey, PA: IGI Global. doi:10.4018/978-1-5225-1046-8.ch011

Senapati, S., & Gupta, V. (2016). Impacts of Climate Change on Fish Productivity: A Quantitative Measurement. In S. Dinda (Ed.), *Handbook of Research on Climate Change Impact on Health and Environmental Sustainability* (pp. 243–260). Hershey, PA: IGI Global. doi:10.4018/978-1-4666-8814-8.ch012

Senapati, S., & Gupta, V. (2016). Vulnerability to Climate Change: Issues and Challenges towards Developing Vulnerability Indicator. In S. Dinda (Ed.), *Handbook of Research on Climate Change Impact on Health and Environmental Sustainability* (pp. 82–104). Hershey, PA: IGI Global. doi:10.4018/978-1-4666-8814-8.ch005

Serban, C., & Maftei, C. (2015). Using Grid Computing and Satellite Remote Sensing in Evapotranspiration Estimation. In C. Maftei (Ed.), *Extreme Weather and Impacts of Climate Change on Water Resources in the Dobrogea Region* (pp. 145–173). Hershey, PA: IGI Global. doi:10.4018/978-1-4666-8438-6.ch006

Shanmuganathan, S., Narayanan, A., & Medagoda, N. P. (2016). Temporal Data Analysis and Mining Methods for Modelling the Climate Change Effects on Malaysia's Oil Palm Yield at Different Regional Scales. In S. Dinda (Ed.), *Handbook of Research on Climate Change Impact on Health and Environmental Sustainability* (pp. 482–513). Hershey, PA: IGI Global. doi:10.4018/978-1-4666-8814-8.ch025

Singh, R. K. (2017). Impact of Climate Change on the Retreat of Himalayan Glaciers and Its Impact on Major River Hydrology: Himalayan Glacier Hydrology. In P. Rao & Y. Patil (Eds.), *Reconsidering the Impact of Climate Change on Global Water Supply, Use, and Management* (pp. 70–83). Hershey, PA: IGI Global. doi:10.4018/978-1-5225-1046-8.ch005

Singh-Ackbarali, D., & Maharaj, R. (2017). Mini Livestock Ranching: Solution to Reducing the Carbon Footprint and Negative Environmental Impacts of Agriculture. In W. Ganpat & W. Isaac (Eds.), *Environmental Sustainability and Climate Change Adaptation Strategies* (pp. 188–212). Hershey, PA: IGI Global. doi:10.4018/978-1-5225-1607-1.ch007

Siwe, A. T., & Tembine, H. (2016). Energy Cost Saving Tips in Distributed Power Networks. In A. Ahmad & N. Hassan (Eds.), *Smart Grid as a Solution for Renewable and Efficient Energy* (pp. 26–48). Hershey, PA: IGI Global. doi:10.4018/978-1-5225-0072-8.ch002

Srivastava, N. (2016). Climate Change Mitigation: Collective Efforts and Responsibly. In S. Dinda (Ed.), *Handbook of Research on Climate Change Impact on Health and Environmental Sustainability* (pp. 427–439). Hershey, PA: IGI Global. doi:10.4018/978-1-4666-8814-8.ch021

Stennikov, V., Barakhtenko, E., Sokolov, D., & Oshchepkova, T. (2016). Problems of Modeling and Optimization of Heat Supply Systems: New Methods and Software for Optimization of Heat Supply System Parameters. In P. Vasant & N. Voropai (Eds.), *Sustaining Power Resources through Energy Optimization and Engineering* (pp. 76–101). Hershey, PA: IGI Global. doi:10.4018/978-1-4666-9755-3.ch004

Stennikov, V., Oshchepkova, T., & Stennikov, N. (2016). Problems of Modeling and Optimization of Heat Supply Systems: Methods to Comprehensively Solve the Problem of Heat Supply System Expansion and Reconstruction. In P. Vasant & N. Voropai (Eds.), *Sustaining Power Resources through Energy Optimization and Engineering* (pp. 26–53). Hershey, PA: IGI Global. doi:10.4018/978-1-4666-9755-3.ch002

Stennikov, V., Penkovskii, A., & Khamisov, O. (2016). Problems of Modeling and Optimization of Heat Supply Systems: Bi-Level Optimization of the Competitive Heat Energy Market. In P. Vasant & N. Voropai (Eds.), *Sustaining Power Resources through Energy Optimization and Engineering* (pp. 54–75). Hershey, PA: IGI Global. doi:10.4018/978-1-4666-9755-3.ch003

Stennikov, V. A., & Postnikov, I. V. (2016). Problems of Modeling and Optimization of Heat Supply Systems: Methodological Support for a Comprehensive Analysis of Fuel and Heat Supply Reliability. In P. Vasant & N. Voropai (Eds.), *Sustaining Power Resources through Energy Optimization and Engineering* (pp. 102–126). Hershey, PA: IGI Global. doi:10.4018/978-1-4666-9755-3.ch005

Stone, R. J. (2017). Modelling the Frequency of Tropical Cyclones in the Lower Caribbean Region. In W. Ganpat & W. Isaac (Eds.), *Environmental Sustainability and Climate Change Adaptation Strategies* (pp. 341–349). Hershey, PA: IGI Global. doi:10.4018/978-1-5225-1607-1.ch013

Swain, M. (2016). Vulnerability to Local Climate Change: Farmers' Perceptions on Trends in Western Odisha, India. In S. Dinda (Ed.), *Handbook of Research on Climate Change Impact on Health and Environmental Sustainability* (pp. 139–155). Hershey, PA: IGI Global. doi:10.4018/978-1-4666-8814-8.ch007

Swain, M., & Swain, M. (2016). Evolution and Efficacy of Drought Management Policies and Programmes: The Case of Western Odisha, India. In S. Dinda (Ed.), *Handbook of Research on Climate Change Impact on Health and Environmental Sustainability* (pp. 176–194). Hershey, PA: IGI Global. doi:10.4018/978-1-4666-8814-8.ch009

Takahashi, Y. (2017). Eco-Friendly On-Site Water Analyses for Ultra-Trace Harmful Ions. In T. Kobayashi (Ed.), *Applied Environmental Materials Science for Sustainability* (pp. 313–326). Hershey, PA: IGI Global. doi:10.4018/978-1-5225-1971-3.ch014

Taokaew, S., Phisalaphong, M., & Newby, B. Z. (2017). Bacterial Cellulose: Biosyntheses, Modifications, and Applications. In T. Kobayashi (Ed.), *Applied Environmental Materials Science for Sustainability* (pp. 255–283). Hershey, PA: IGI Global. doi:10.4018/978-1-5225-1971-3.ch012

Tauch, S., Liu, W., & Pears, R. (2016). Measuring Cascading Failures in Smart Grid Networks. In A. Ahmad & N. Hassan (Eds.), *Smart Grid as a Solution for Renewable and Efficient Energy* (pp. 208–225). Hershey, PA: IGI Global. doi:10.4018/978-1-5225-0072-8.ch009

Trang, T. T. (2017). Study on the Use of Biomass Polymer Sheets in Water/Alcohol Pervaporation as a Sustainable Source of Alcohol Energy. In T. Kobayashi (Ed.), *Applied Environmental Materials Science for Sustainability* (pp. 234–254). Hershey, PA: IGI Global. doi:10.4018/978-1-5225-1971-3.ch011

Tsiaras, S. (2017). Exploring the Impact of Tourism to the Sustainable Development of Mountain Regions: Implications of the Climatic Conditions. *International Journal of Agricultural and Environmental Information Systems*, *8*(1), 14–28. doi:10.4018/IJAEIS.2017010102

Tugjamba, N., Yembuu, B., Gantumur, A., & Getsel, U. (2016). Policy Provisions and Teachers' Needs on Climate Change Education for Sustainable Development in Mongolia. *International Journal of Asian Business and Information Management*, *7*(4), 36–48. doi:10.4018/IJABIM.2016100103

Tyukhov, I., Rezk, H., & Vasant, P. (2016). Modern Optimization Algorithms and Applications in Solar Photovoltaic Engineering. In P. Vasant & N. Voropai (Eds.), *Sustaining Power Resources through Energy Optimization and Engineering* (pp. 390–445). Hershey, PA: IGI Global. doi:10.4018/978-1-4666-9755-3.ch016

Uddin, Z., Shah, N., Ahmad, A., Mehmood, W., & Alam, F. (2016). Signal Processing Techniques in Smart Grids. In A. Ahmad & N. Hassan (Eds.), *Smart Grid as a Solution for Renewable and Efficient Energy* (pp. 273–297). Hershey, PA: IGI Global. doi:10.4018/978-1-5225-0072-8.ch012

Ventrapragada, E. A., & Rayavarapu, N. (2017). Climate Change and Agriculture: Time for a Responsive and Responsible System of Water Management. In P. Rao & Y. Patil (Eds.), *Reconsidering the Impact of Climate Change on Global Water Supply, Use, and Management* (pp. 326–363). Hershey, PA: IGI Global. doi:10.4018/978-1-5225-1046-8.ch018

Wang, Q., & Liu, H. (2017). Optimized Base Station Sleeping and Renewable Energy Procurement Scheme Using PSO. *International Journal of Swarm Intelligence Research*, *8*(1), 54–73. doi:10.4018/IJSIR.2017010103

Wong, J., & Lim, Y. S. (2016). Revolution of Energy Storage System in Smart Grids. In A. Ahmad & N. Hassan (Eds.), *Smart Grid as a Solution for Renewable and Efficient Energy* (pp. 181–206). Hershey, PA: IGI Global. doi:10.4018/978-1-5225-0072-8.ch008

Wu, Y., Tan, X., Qian, L. P., Tsang, D. H., Song, W., & Yu, L. (2016). Management of Scheduling and Trading in Hybrid Energy Trading Market. In A. Ahmad & N. Hassan (Eds.), *Smart Grid as a Solution for Renewable and Efficient Energy* (pp. 123–148). Hershey, PA: IGI Global. doi:10.4018/978-1-5225-0072-8.ch006

Xiang, M., Liu, W., Bai, Q., & Al-Anbuky, A. (2016). Dynamic Trust Elective Geo Routing to Secure Smart Grid Communication Networks. In A. Ahmad & N. Hassan (Eds.), *Smart Grid as a Solution for Renewable and Efficient Energy* (pp. 323–343). Hershey, PA: IGI Global. doi:10.4018/978-1-5225-0072-8.ch014

Zăgan, S., & Chiţu, M. (2015). The Influence of Air Temperature on the Quality Parameters of the Black Sea Coastal Waters. In C. Maftei (Ed.), *Extreme Weather and Impacts of Climate Change on Water Resources in the Dobrogea Region* (pp. 174–204). Hershey, PA: IGI Global. doi:10.4018/978-1-4666-8438-6.ch007

Zharkov, S. (2016). Assessment and Enhancement of the Energy Supply System Efficiency with Emphasis on the Cogeneration and Renewable as Main Directions for Fuel Saving. In P. Vasant & N. Voropai (Eds.), *Sustaining Power Resources through Energy Optimization and Engineering* (pp. 1–25). Hershey, PA: IGI Global. doi:10.4018/978-1-4666-9755-3.ch001

About the Author

Hiroshi Ueda, 1971-1975: Faculty of Fisheries, Hokkaido University, Hakodate. 1975-1980: Graduate School of Fisheries Science, Hokkaido University, Hakodate. 1980-1983: Postdoctoral Fellowship supported by Japanese Society of Promotion of Science (JSPS), National Institute for Basic Biology, Okazaki. 1983-1991: Assistant Professor, University of Occupational and Environmental Health, School of Medicine. Kitakyushu. 1985-1986: Visiting Scientist, Worcester Foundation of Experimental Biology. Worcester, USA. 1986-1987: Forgaty Fellowship supported by National Institute of Health (NIH), University of North Carolina, Chapel Hill, USA. 1991-2001: Associate Professor, Faculty of Fisheries, Toya Lake Station for Environmental Biology, Hokkaido University, Abuta. 2001-2015: Professor, Field Science Center for Northern Biosphere and Graduate School of Environmental Science, Hokkaido Univerity, Sapporo. 2015-2017: Specially Appointed Professor, Field Science Center for Northern Biosphere and Graduate School of Environmental Science, Hokkaido University, Sapporo. 2017-Present: Professor Emeritus, Hokkaido University. Technical Advisor, Hokkaido Aquaculture Promotion Corporation, Sapporo.

Index

D

E

F

G

H

I

K

L

M

N

O

U

V

W

www.ingramcontent.com/pod-product-compliance
Ingram Content Group UK Ltd.
Pitfield, Milton Keynes, MK11 3LW, UK
UKHW050945290926
15470UKWH00005B/1640

9 781799 820543